Pepsi Memorabilia Then and Now

An Unauthorized Handbook and Price Guide

Phil Dillman
& Larry Woestman

4880 Lower Valley Road, Atglen, PA 19310 USA

Dedication

To the families of Phil and Larry for their patience and tolerance of a passion that exceeds that of a typical collector, and to the many Pepsi collectors that have become our friends for life.

Title Page and Back Cover: Artwork courtesy of Randy Schwentker

Pepsi-Cola and all of the flavors listed in the "flavors" chapter are registered trademarks of Pepsico, Inc. All of the information contained herein was researched by the authors through reading, conversations, and their membership in the Pepsi-Cola Collectors Club.

Library of Congress Catalog Card Number: 99-69800

Book Design by Anne Davidsen
Type set in Impress /Zurich

ISBN: 0-7643-1105-0

Printed in China
1 2 3 4

Published by Schiffer Publishing Ltd.
4880 Lower Valley Road
Atglen, PA 19310
Phone: (610) 593-1777; Fax: (610) 593-2002
E-mail: Schifferbk@aol.com
Please visit our web site catalog at
www.schifferbooks.com
We are always looking for people to write books on new and related subjects. If you have an idea for a book, please contact us at the above address.

This book may be purchased from the publisher.
Include $3.95 for shipping.
Please try your bookstore first.
You may write for a free catalog.

In Europe, Schiffer books are distributed by:
Bushwood Books
6 Marksbury Ave.
Kew Gardens
Surrey TW9 4JF England
Phone: 44 (0)208 392-8585; Fax: 44 (0)208 392-9876
E-mail: Bushwd@aol.com
Free postage in the UK. Europe: air mail at cost.
Try your bookstore first.

Contents

Acknowledgments	4
Preface	4
Important Information	5
Pepsi Flavors	7
Odds and Ends	9
Audio/Visual	10
Awards	13
Belt Buckles	16
Books and Magazines	18
Bottle Openers	21
Bottles	22
Cans	30
Carriers	38
Clocks	49
Clothing	53
Commemorative Items—100th Anniversary	57
Coolers	68
Desk Items	73
Drinking Glasses	76
Food Related	78
Fountain Related	83
Jewelry	85
Knives	87
Lights	88
Miscellaneous	89
Paper Items	96
Pencils and Pens	103
Pinback Buttons, Patches, and Hatpins	103
Plates	106
Point-of-Purchase (P.O.P.)	107
Radios	112
Santas	113
Signs	115
Smoking Related	124
Thermometers	127
Toys	128
Toy Trains	133
Toy Trucks	135
Trays	137
Umbrellas	138
Vending Machines	140
Watches	141

Acknowledgments

This book would not be nearly as interesting if not for the input of the following people:

Roger and Janet Blad
Dick Bridgforth
Wayne Burgess
Weylin Buzby
Chris Dimitt
John Hantz
John D. Kennedy
Russ and Betty Kimbler
Kaye LeMahieu
Mike Noll
Randy Schwentker
Brian Sheeler
Ron and Kathy Stines
Bob Stoddard
Lenny and Laura Vigna
Karen Weaver

In Memory of Richard (Dick) Kehn

Preface

If this is your first book on Pepsi collectibles—Welcome! Currently, we know of six books already available regarding Pepsi collectibles, not including the books that just cover bottles. Since each of these books are meant as guides to help Pepsi collectors date and value their pieces, and because these books are fairly simple to use, we felt it would be helpful to follow basically the same guidelines as previously established. However, not included in these guides were many pieces that had yet to be catalogued. Therefore, you will find very few of the items in our book in any of the other books. We also believe there are plenty of other recent items out there that should be considered "collectibles" due to their limited production or distribution; many of those items are included here as well. We think most of the users of this guide will appreciate the inclusion and history of the many different flavors sold by Pepsi during their first 100 years in existence.

There is only one abbreviation in this book that we need to explain, ACL—Applied Color Label—which refers to the labels painted on bottles.

Having said all that, we could each use a nice, cold Pepsi. Care to join us?

Important Information

Collectors Club

The Pepsi-Collectors Club offers a great forum for the buying, selling, and trading of Pepsi collectibles. Members receive a bi-monthly newsletter, information about annual Pepsi Fest events, and opportunities to acquire special limited edition club commemorative items, plus the ability to network with collectors nationwide and internationally. If intersted in membership, please write to the below address for a Membership Application.

Pepsi-Cola Collectors Club
P.O. Box 817
Claremont, CA 91711

Pepsi's 100th Anniversary

As most Pepsi collectors already know, 1998 marked the 100th anniversary of the drink Pepsi-Cola. Many of the items shown in this book were only available at the 100th anniversary celebration in New Bern, North Carolina, and some of these items were extremely limited in production. Since these items are so new, the value shown is the amount paid for each item in 1998.

Reproduction and Fantasy Items

With all of the reproduction items appearing at various shows and flea markets lately, we need to address this issue.

Reproductions are copies of original items. Some of these items include the 1909 watch fob with U6705 on the back, the 27-inch aluminum bottle cap sign, and the die-cut aluminum bottle sign, each being re-issued by Stout Sign Company. There are also many 1940s cardboard signs being reproduced.

Fantasy items are those created using old logos to make an item that previously never existed. The best example of this is the red thermometer with a bottle cap logo at the top and bottom. THE RED THERMOMETERS ARE NOT OLD. They were produced during the late 1980s and early 1990s by Taylor Environmental Instruments out of Fletcher, North Carolina, and assembled in Mexico. Another item would be the rectangular sign with the 1950s bottle, a yellow background, and 5c. None of those three items belong together on the same sign.

Of course, it's okay to have these items in your collection as long as you understand that they are not originals. If you aren't sure if an item is an original, call a fellow collector and ask if they are familiar with the item. Otherwise, have the dealer guarantee the item with the option of returning it.

For further examples of reproductions and fantasy items, refer to some of the Pepsi collectibles web sites on the Internet.

Test and Prototype Items

We are certain that not everyone will agree with the values assigned to the test or prototype pieces; however, there are so few of any of these pieces around that you should expect to pay a substantial amount for these items, if you can find them. Many of these type items are listed in this book with the word "RARE" in place of the price. Based on what collectors have paid for some of these items, and only as a general reference, the values of these pieces can range from $200.00 each to more than $1,000.00 each.

Pepsi Stuff

The Pepsi Stuff promotion began in 1996 and created the opportunity for the packaging on Pepsi and Diet Pepsi to be saved for points, which could in turn be redeemed for free "stuff" such as Pepsi T-shirts, hats, phone cards, patches, beach chairs, duffel bags, etc. In 1997, Pepsico also included Mountain Dew "stuff," with points printed on Mountain Dew and Diet Mountain Dew packaging as well as Pepsi and Diet Pepsi packaging.

Mountain Dew

The following information was provided by Dick Bridgforth and Wayne Burgess.

Mountain Dew is one of the most successful flavors in Pepsi-Cola' s line-up. The sales of Mountain Dew have seen phenomenal growth since it was purchased by Pepsi. Because of its popularity, the second year of the "Pepsi Stuff" promotion included Mountain Dew items as well.

Many people include Mountain Dew in their Pepsi collecting, with the most sought after items being those that picture the hillbilly. This also includes the early Mountain Dew bottles, the first of which was produced in 1951. What many people don't realize is that the original flavor of Mountain Dew actually tasted like 7-Up. It was reformulated in 1962 to its current flavor. Then, on September 2, 1964, The Pepsi-Cola Company purchased all of the capital stock of Tip Corp., owner of Mountain Dew. Tip continued to market Mountain Dew as a wholly-owned subsidiary of Pepsico. On December 30, 1965, Pepsico acquired all of the assets of Tip, including the trademark "Mountain Dew," and the rest is history.

Name Bottles

Mountain Dew name bottles came in seven sizes. Of the approximately 1000 known, early bottles, most are the 10-ounce size. Other sizes include 7, 8, 8 ¾, 9, 12, 16, and 24 ounce. All of the bottles are green except for the very first "by BARNEY and ALLY" bottles, which are clear.

As of fall of 1999, 721 unique Mountain Dew name bottles had been catalogued with various names on them. Early Mountain Dew bottles included the names of salemen, managers, bottling plant owners, etc. The majority of the bottles (550) have one or two names on them. Only a few (14) have five or more names. Other bottles (71) feature the names of towns or counties. There is even one named for Good Time Charlie, a dog.

Since this list is always growing as new name bottles are discovered, it is best to check out the Mountain Dew Bottle Collectors Home Page on the Internet from time to time.

Pepsi Flavors

While it is true that the main concern of Pepsico and their competitors are their cola drinks, the fact is, not everyone likes cola. Even those that do like cola sometimes prefer a change. That's why most cola makers also provide a line of flavored drinks. By offering their customers a choice, the companies increase both their overall beverage per capita sales numbers and market share. They also increase their chances for customer brand loyalty. For example, a Pepsi drinker who desires a root beer will probably choose Mug, since it is made by the same company and, moreover, it is usually sold in close proximity to Pepsi. Obviously, this isn't always the case; but with the right advertising and product placement, brand loyalty increases. Since this is true for every soda company, the "cola wars" have escalated into the "beverage wars." To this end, the market has been saturated with more flavors than you would believe, with variations of existing flavors, combined flavors, and flavors never before imagined. Many early soda bottlers actually added Pepsi-Cola to their line-up as simply one more flavor to sell. Since cola drinks were still fairly new, other flavors such as orange and grape were the principle beverages. As colas became more popular, however, flavors took a back seat and cola became the priority.

Evervess, a sparkling water, could be considered Pepsi-Cola's first "flavor;" introduced in 1946, it could be used as a drink mixer as well as a beverage worth drinking by itself. **Teem**, however, would have to be the company's first true flavor, a lemon-lime drink introduced in 1959.

In 1960, the **Patio** brand was created to market various flavors such as **orange**, **grape**, **root beer**, **strawberry**, **ginger ale**, **club soda**, **red cream**, **red cherry**, **grapefruit**, and **tonic water**. 1963 saw the test-marketing of **Patio Diet Cola**. Although sales of this flavor were respectable, The Pepsi-Cola Company could not comfortably afford the cost of advertising for additional products. So partly out of necessity, in 1964, the name of Patio Diet Cola was changed to **Diet Pepsi-Cola** and was advertised right alongside Pepsi. 1964 also saw the addition of **Mountain Dew** to the line-up, along with the test-marketing of **Sugar Free Teem**.

Other experimental flavors were test-marketed during the 1960s: with **Devil Shake**, a chocolate drink in 1966; **Tropic Surf**, a clear, dietary citrus beverage, in 1967; **Pepsi Snoball**, a frozen sippin' ice, also in 1967; and **Skandi**, a lemon-flavored diet cola, again, in 1967.

It was also during this time that John Sculley, president of Pepsi-Cola from 1977 to 1983, developed some dramatically different flavors with an Oriental motif, such as mandarin orange and cherry blossom. These flavors were certainly innovative and very much ahead of their time. And, while these flavors had possibilities, Sculley, and others working on this project, learned that the other cola company from Atlanta had somehow obtained drawings of these products and their formulas; thus, the entire project was scrapped.

In the mid-1970s, **Sugar Free Mountain Dew** was test-marketed. The next entry into this category was **Pepsi Light**. Introduced in 1976, it was Diet Pepsi with a twist of lemon. It went national in 1979. **On-Tap Root Beer** began its test-market in Milwaukee in August of 1977. A 1978 test market flavor was **Aspen**. Tested in Los Angeles, this was a clear soda "with a snap of apple."

Pepsi Free and **Sugar Free Pepsi Free** made their debut in 1981 and went national in 1983 to help fulfill a public demand for caffeine-free drinks. **Lemon Lime Slice** and **Diet Lemon Lime Slice** were added in 1984, each flavor boasting 10% real fruit juices.

1986 was a busy year for Pepsi with the addition of **Mandarin Orange Slice**, **Diet Mandarin Orange Slice**, **Apple Slice**, **Diet Apple Slice**, **Cherry Cola Slice**, and **Diet Cherry Cola Slice**. Also in 1986, Pepsico purchased **Mug Root Beer** and **Diet Mug Root Beer** to replace On-Tap. **Diet Mountain Dew** hit the scene in 1987. Also in 1987, **Pepsi A.M.** and **Diet Pepsi A.M.** (though unsuccessful in the attempt) addressed the people who prefer to drink Pepsi in the morning. These drinks contained more caffeine and less carbonation. **Jake's Diet Cola** was test-marketed in 1987 in Jacksonville, Seattle, and Indianapolis.

Wild Cherry Pepsi and **Diet Wild Cherry Pepsi** replaced Cherry Cola Slice and Diet Cherry Cola Slice in 1988. That same year, **Mountain Dew Red** and **Diet Mountain Dew Red** were test-marketed in Birmingham, Alabama, but never made it past the test-market stage. Also in 1988, Pepsi Free and Diet Pepsi Free were renamed **Caffeine Free Pepsi** and **Caffeine Free Diet Pepsi**.

Another variation on Mountain Dew was tested in 1989 and called **Mountain Dew Sport**. There was also a two-calorie version called **Mountain Dew Diet Sport**. Although these flavors realized only modest success, they helped set the stage for a new line of Pepsi products known as "All Sport."

The **H2OH!** line of flavors was introduced in 1989 and consisted of **Lemon Lime**, **Berry Splash**, **Orange**, and **Sparkling Water**.

Another flavor variation came with the 1989 reformulation of Orange Slice and Lemon Lime Slice. The 10% real fruit juices were dropped and the package graphics were changed. Two more flavors added in 1989 were **Mug Cream Soda** and **Diet Mug Cream Soda**.

The 1991 introduction of the "Wild Ones" provided an interesting new twist to Pepsi: **Raging Razzberry Pepsi**, **Diet Raging Razzberry Pepsi**, **Strawberry Burst Pepsi**, **Tropical Chill Pepsi**, and **Diet Tropical Chill Pepsi**. According to a Pepsi publication, Raging Razzberry was tested in Peoria, Illinois, and in Sacramento, California, along with Strawberry Burst and Tropical Chill; while all three flavors and their diet counterparts were tested in Tulsa, Oklahoma. As of this book's publication, a diet version of Strawberry Burst has yet to be verified.

All Sport isotonic drinks were introduced in 1991. As of 1999, these flavors included **Lemon Lime**, **Grape**, **Orange**, **Fruit Punch**, **Blue Ice**, **Cherry Slam**, **Raspberry Burst**, and **Extreme Watermelon**.

In 1992, one of the most innovative drinks of the decade was welcomed with great fanfare and extensive media coverage, the clear cola drink called **Crystal Pepsi**. Along with **Diet Crystal Pepsi**, these colas answered a new consumer demand for purity. Unfortunately, these drinks didn't taste like Pepsi, which most folks had expected, and, after a little more than a year, Crystal Pepsi was reformulated. Also during 1992, **Grape Slice** and **Strawberry Slice** were introduced to the public.

Pineapple Slice was introduced in 1993.

Caffeine Free Mountain Dew and **Caffeine Free Diet Mountain Dew** were added in 1994. Also in 1994, Pepsi joined the growing bottled water market with **Aquafina**. The reformulated version of Crystal Pepsi was released as **Crystal...from the makers of Pepsi**, available only in regular. Sadly, it realized even less success than Crystal Pepsi.

A U.S. version of Pepsi Max was tested in Florida under the name **Pepsi XL** (excellent taste, less sugar) during April of 1995. Again, the test market was the end of the line for this particular flavor. **Josta** also hit the scene in 1995. It was a very sweet, highly carbonated beverage made with Guarana berries (which in some countries are considered an aphrodisiac) and lots of caffeine. (It lasted until 1999.)

Pepsi Kona Coffee Cola and **Diet Pepsi Kona Coffee Cola** were tested in Philadelphia in May and June of 1996. This unique cola drink combined the tastes of Pepsi and coffee beans grown in Kona, Hawaii.

1998 saw the beginning of two new Pepsi products: **Storm**, a lemon-lime drink with caffeine and virtually no aftertaste, and **Pepsi One**, a one-calorie cola designed to taste more like regular Pepsi than Diet Pepsi.

A diet version of Storm called **Light Storm** was released in 1999.

Pepsico also has a few flavors that are only available outside of the United States. Introduced in 1957, **Mirinda** remains a very successful orange drink for Pepsi-Cola International. In 1992, a West Malaysia Pepsi bottler introduced four new Mirinda flavors: **Grape**, **Pineapple**, **Krim V**, and **Apple**.

1993 saw the International release of **Pepsi Max**. This cola tasted similar to Pepsi but has only 1/3 of the calories. It has not been sold in the U.S. as it contains a sweetener not approved by the U.S.F.D.A. (United States Food and Drug Administration).

Kas is a line of flavored drinks that was acquired from Spain in 1993.

Pepsico owns **7-Up** worldwide, except in the United States, through a 1986 purchase. Pepsico has also distributed original and/or reformulated versions of **Evervess**, **Skandi**, and **Teem** throughout various countries outside of the U.S. at different times during the 1970s, '80s, and '90s.

Back in the U.S., during the late 1980s, Pepsico decided to revive Pepsi Sno-Ball, a slush-type drink, under the name **Pepsicle**, available in both Pepsi and Mountain Dew flavors. However, due to legal challenges against the use of that name, the flavors were released during the early 1990s with the names **Pepsi Freeze** and **Mountain Dew Freeze**.

One oddity in the world of flavors is the non-drink items. In 1998, there was a gumball machine-type of candy called Slice that was marketed by Pepsico.

We have listed all of the flavors with which we are familiar. Other flavors that we know to exist, but have little information on, are listed below. Please let us know of any flavors we may have missed or are unaware of.

Cherry Pepsi (Canada)
Lemon Pepsi (possible predecessor of Pepsi Light)
Dr. Slice (U.S.)
Strawberry Pepsi and **Tropical Pepsi** (England)
Schwip Schwap (Germany)
Paso de las Toros

Odds and Ends

The Crystal tank car #4205 listed as "also available" on the New Pepsi Generation train set was never produced.

There were a number of items put out by Pepsi that incorrectly match the year of 1940 with the single dot script logo, which did not exist until 1951.

The book *Twelve Full Ounces* was printed first in 1962, and again in 1964, with the latter version including an addendum.

Audio/Visual

001 Training record, c.1940s, 16", $25.00

002 Record, *Rationing*, c.1940s, 12", $20.00

003 Record, sound effects, 1960, 10", $20.00

004 Record, c.1980s, 7", $10.00

005 Record, Christmas music, c.1960s, 12", $10.00

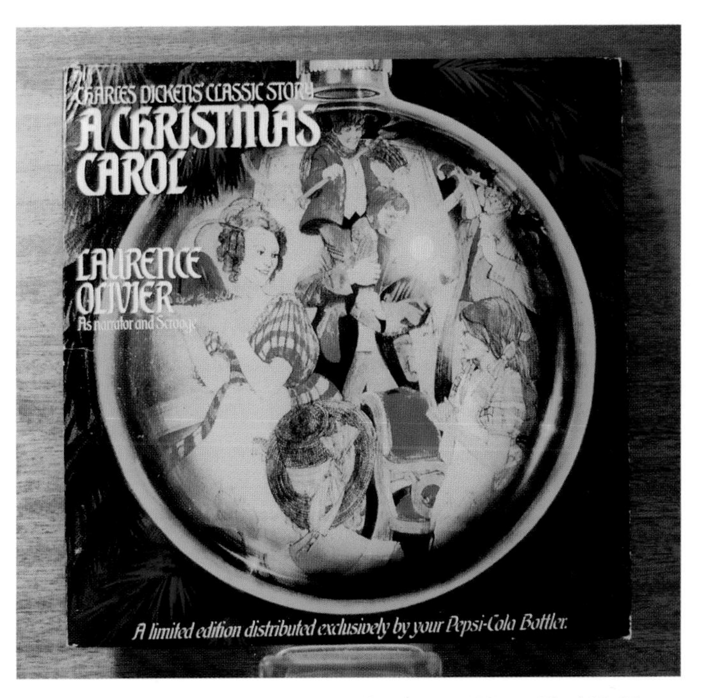

006 Record, Story—*A Christmas Carol*, c.1970s, 12", $10.00

007 Record, radio broadcast of *Superman*, c.1970s, 12", $10.00

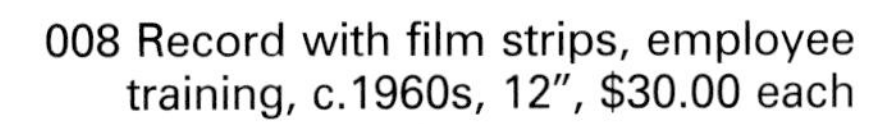

008 Record with film strips, employee training, c.1960s, 12", $30.00 each

009 Record, commercials, 1977, 12", $20.00

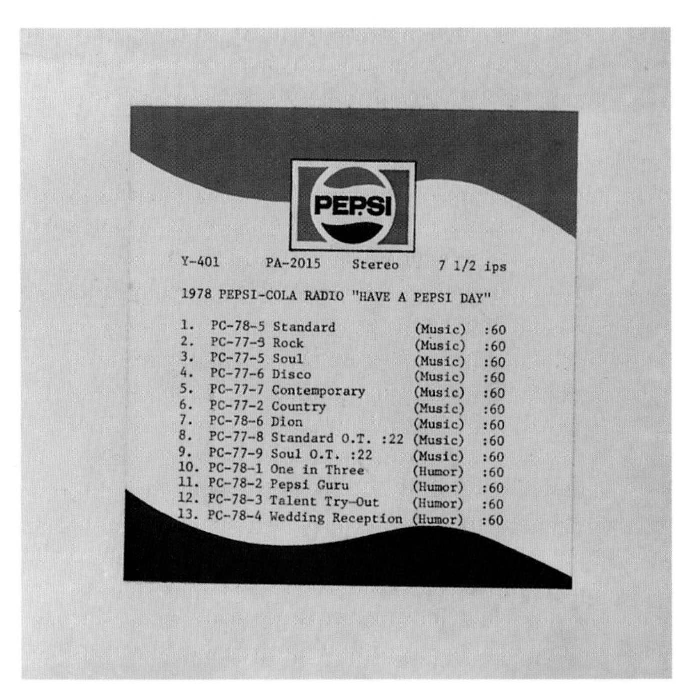

011 Reel to reel tape, commercials, 1978, 7", $5.00

010 Record, commercials, 1981, 12", $20.00

012 Video tape, movie, c.1990s, V H S, $5.00

013 Video tape, bottler's preview, 1992, VHS, $5.00

014 Video tape, commercials, 1986, 3/4" format, $10.00

015 Mini record player and record card, 1980, $150.00

Awards

016 Plant opening plaque, brass on wood, 1961, 13.75" x 17.75", $250.00

017 Travel award, ceramic tile and wood, 1962, 10" x 10", $30.00

018 Training award, P C M I, metal on wood, c.1960s, 14.5" x 8.5", $20.00

022 President's award for per capita, metal on wood, 1979, 20", $150.00

019 Route Salesman of the Month award, metal on wood, 1972, 4" x 8", $30.00

021 Trainer's award, metal on wood, 1977, 6" x 10", $30.00

020 Star Performer award, plastic on wood, c.1970s, 9.5" x 8.75", $15.00

023 Safe Driver award, metal on wood, 1986, 12" x 9", $15.00

024 Effie award for advertising placement, metal on wood, 1988, 11" x 15", $15.00

025 Fountain Service plaque, wood, c.1980s, 8" x 10", $10.00

027 Years of Service and Dedication plaque, metal on wood, 1991, 12" x 9", $15.00

026 Merchandiser of the Month plaque, wood, c.1980s, 5" x 7", $10.00

Belt Buckles

028 Bookend logo, brass, c.1980s, $5.00

029 Right One, Baby, brass, 1993, $10.00

030 Bookend logo in white flower, metal and enamel, c.1980s, $10.00

031 Memphis, Missouri bottling plant, metal, 1983, $20.00

032 Bookend logo, metal with enamel, c.1970s, $5.00

033 8-pack of 16 oz. bottles, metal, c.1980s, $8.00

034 Single dot script with ornate edge, brass, c.1990s, $7.00

035 Round, vinyl on metal, c.1970s, $5.00

036 Single dot script with "Hits the Spot," brass, c.1970s, $10.00

037 Single dot script with steer horns above, metal, c.1980s, $10.00

038 Bottle cap with single dot script, brass, c.1960s, $10.00

039 Square with Pepsi Challenge Pot of Gold, brass, 1986, $10.00

040 Coho '79 Chicago Park District, vinyl on metal, 1979, $10.00

041 Boxed set of two buckles, brass, 1986, $30.00

042 Mountain Dew, Hello Sunshine, brass, c.1980s, $10.00

Books and Magazines

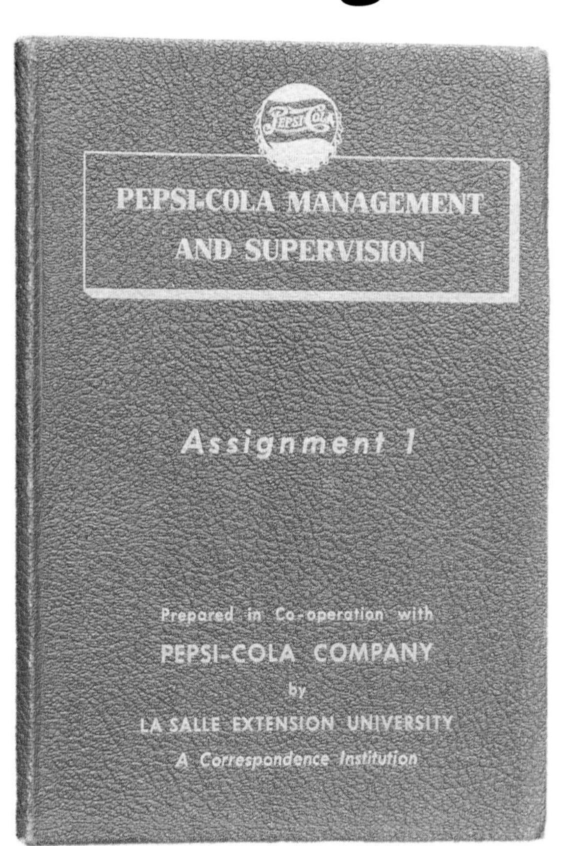

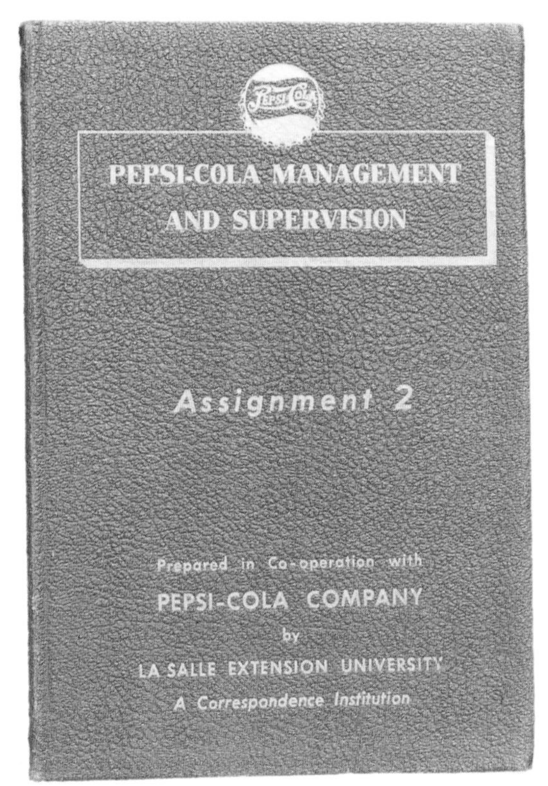

Above: 044 Management Training, 1947, 6.75" x 9.75", $100.00

Left: 043 Management Training, 1947, 6.75" x 9.75", $100.00

045 Production School 3-ring binder, 1959, $30.00

046 Quality Control 3-ring binder, c.1960s, $75.00

047 *Twelve Full Ounces*, with dust jacket, 1962, $70.00

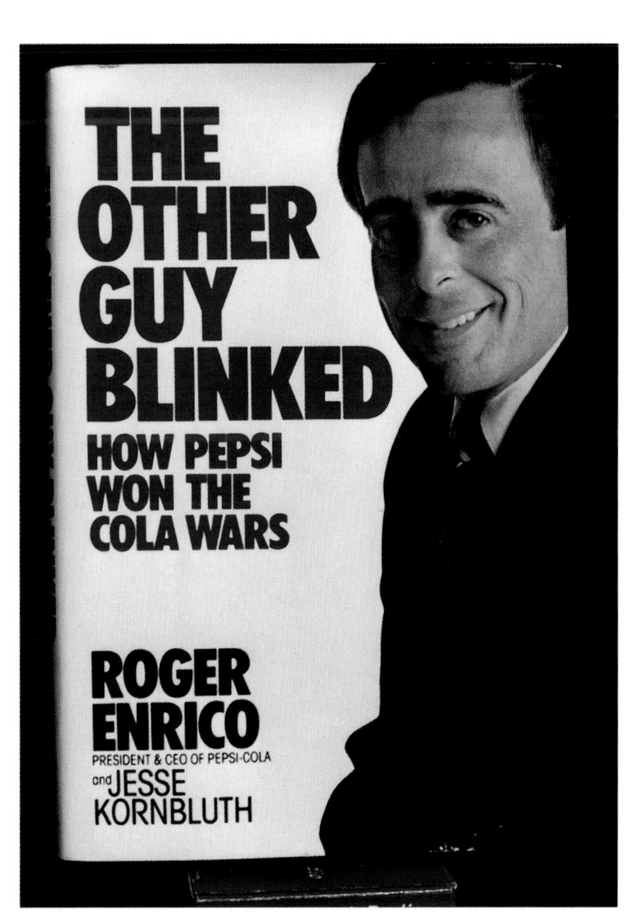

048 *The Other Guy Blinked*, with dust jacket, 1986, $15.00

049 *A Pepsi Day*, soft cover, 1990, $20.00

050 *People in General* magazines, leather bound set, 1988-1996, $50.00

051 Magazine, 1963, $10.00

052 Magazine, 1977, $5.00

053 Magazine, Pepsi's 100th anniversary issue, 1998, $10.00

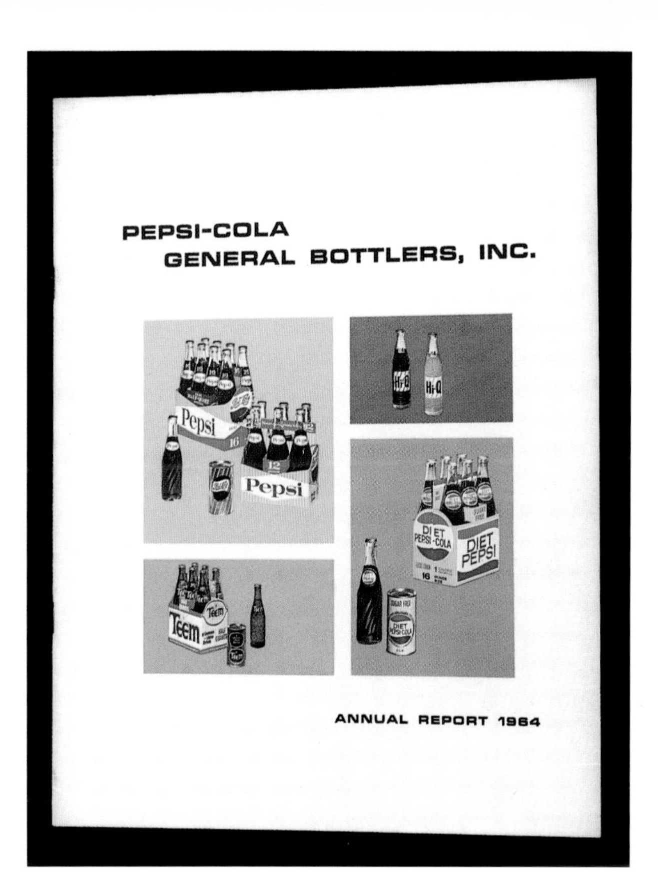

054 Annual Report, 1964, $10.00

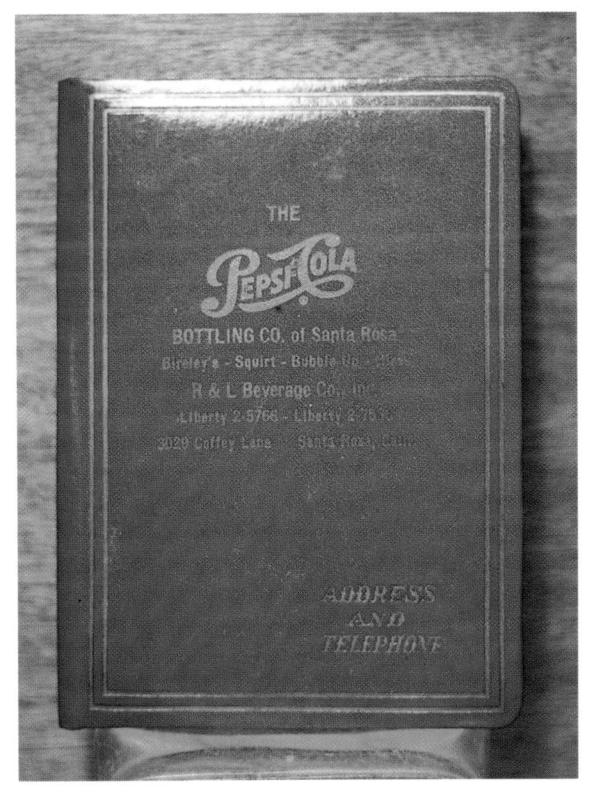

055 Address book, single dot script, c.1950s, $10.00

Bottle Openers

056 Wire type, block letters, c.1960s, $5.00

057 Bottle shaped, metal, c.1940s, $35.00

058 Wall mount, brass, c.1960s, $15.00

059 Evervess, metal, c.1940s, $50.00

060 Bottle shaped, wood handle with wire top, c.1980s, $10.00

Bottles

061 Glass, acid etched, Seltzer bottle, c.1920s, $900.00

062 Amber glass with foil label, c.1960s, 16 oz., $35.00

063 Clear glass, A C L, c.1960s, 10 oz., $35.00

064 Clear glass with embossed large dimples, test bottle, c.1960s, 10 oz., RARE

065 Clear glass with etched swirls, test bottle, c.1960s, 16 oz., RARE

From left: 066 Light green glass, A C L, c.1970's, 32 oz., $50.00
067 Dark green glass, A C L, Diet Pepsi, c.1970s, 16 oz., $50.00
068 Dark green glass, A C L, straight sided, c.1960s, 16 oz., $50.00
069 Dark green glass, A C L, c.1960s, 12 oz., $50.00
070 Medium green glass, A C L, c.1970s, 12 oz., $50.00
071 Medium green glass, A C L, c.1980s, 12 oz., $50.00

From left: 072 Medium green glass, A C L, c.1960s, 12 oz., $50.00
073 Medium green glass, no label, c.1960s, 12 oz., $40.00
074 Dark green glass, A C L, Diet Pepsi, c.1970s, 12 oz., $50.00
075 Dark green glass, A C L, Pepsi Free, c.1980s, 12 oz., $50.00
076 Dark green glass, A C L, Sugar Free Pepsi Free, c.1980s, 12 oz., $50.00

From left: 077 Medium green glass, embossed, c.1970s, 10 oz., $50.00
078 Dark brown glass, embossed, c.1970s, 10 oz., $50.00
079 Dark brown glass, embossed, c.1960s, 16 oz., $50.00
080 Light blue glass, embossed, c.1960s, 10 oz., $40.00

081 Plastic, first 2-liter bottle, Munster, Indiana, 1977, $75.00

082 Clear glass with styro-wrap, plant opening, Winston- Salem, 1984, $20.00

083 Longneck, A C L, set of four, Richard Petty, 1991, 12 oz., $25.00

084 Longneck, A C L, Duke Blue Devils, 1991, 12 oz., $75.00

085 Longneck, A C L, Cindy Crawford, 1992, 12 oz., $75.00

086 Longneck, A C L, Fred Couples, 1992, 12 oz., $75.00

087 Longneck, A C L, Terry Bradshaw, 1992, 12 oz., $75.00

088 Longneck, A C L, Diet Pepsi, Ray Charles, 1992, 12 oz., $75.00

089 Longneck, A C L, set of eight, Richard Petty, 1992, 12 oz., with case, $100.00

090 Plastic, The Grip, test bottle, 1998, 2 liter, $5.00

091 Glass, A C L, Diet Pepsi, c.1960s, 10 oz., $25.00

092 Glass with paper label, Diet Pepsi, c.1970s, 7 oz., $20.00

093 Glass with foil label, Diet Pepsi, c.1970s, 10 oz., $15.00

Left: 094 Plastic with plastic wrap, Pepsi A. M., test item, 1987, 16 oz., $10.00

Right: 095 Plastic with plastic wrap, Diet Pepsi A. M., test item, 1987, 16 oz., $10.00

096 Glass with plastic wrap, Crystal Pepsi, 1992, 16 oz., $5.00

097 Plastic with plastic wrap, Crystal, test item, 1994, 20 oz., $5.00

098 Plastic with plastic wrap, Pepsi Kona Coffee Cola, test item, 1996, 20 oz., $3.00

Left: 099 Glass, A C L, Lemon Pepsi, test item?, c.1970s, 10 oz., $15.00
Right: 100 Glass with styro-wrap, Lemon Pepsi, test item?, c.1980s, $15.00

101 Glass, A C L, Pepsi Light, c.1970s, 32 oz., $15.00

102 Plastic with plastic wrap, Pepsi X L, test item, 1995, $5.00

103 Glass, A C L, Evervess, c.1940s, 32 oz., RARE

104 Glass, A C L, Devil Shake, c.1960s, 10 oz., RARE

105-110 Group of six glass H2 Oh! bottles (various color paper labels shown), 1989, 10 oz. and 20 oz. size, $5.00 each

111 Glass, A C L, Jake's Diet Cola, 1987, 16 oz., $10.00

112 Glass, embossed, Mountain Dew, c.1960s, 10 oz., $10.00

113 Glass, A C L, Mountain Dew, by Charlie and Bill, c.1950s, 24 oz., RARE

115-117 Three glass with styro-wrap bottles, Mountain Dew Sport and Mountain Dew Diet Sport, test items, 1989, 16 oz., $5.00 each

114 Glass, A C L, Mountain Dew Party Jug, test item?, c.1960s, 32 oz., RARE

118 Glass with paper label, Patio, c.1960s, 10 oz., $30.00

119 Glass, embossed, Patio, c.1960s, 10 oz., $30.00

120 Clear glass, A C L, Skandi, test item, c.1960s, 10 oz., RARE
121 Light blue glass, A C L, Skandi, test item, c.1960s, 10 oz., RARE
122 Glass with foil label, Skandi, test item, c.1960s, 16 oz., RARE

123 Green glass with paper label, Teem, c.1960s, 32 oz., $40.00

124 Green glass, A C L, Sugar Free Teem, test item, 1964, 12 oz., $75.00

125 Green glass, A C L, Tropic Surf, 1967, 10 oz., $75.00

126 Green glass, A C L, Tropic Surf, 1967, 10 oz., $75.00

Cans

127 Steel, c.1960s, 16 oz., RARE

128 Aluminum, Diet Pepsi, c.1970s, 6.3 oz., $10.00

129 Aluminum, Diet Pepsi Free, c.1980s, 6.3 oz., $10.00

130 Aluminum, set of five, A. J. Foyt, c.1980s, 12 oz., $25.00

From left: 132 Aluminum, Pepsi Slam Can, 1995, 16 oz., $5.00
133 Aluminum, Diet Pepsi Slam Can, 1995, 16 oz., $5.00
134 Aluminum, Mountain Dew Extreme Can, 1995, 16 oz., $5.00

131 Aluminum, Grand Opening, St. Louis, Missouri, 1994, 12 oz., $10.00

From left: 135 Aluminum, Pepsi, test market small globe, 1997, 12 oz., $5.00
136 Aluminum, Caffeine Free Pepsi, test market small globe, 1997, 12 oz., $5.00
137 Aluminum, Diet Pepsi, test market small globe, 1997, 12 oz., $5.00
138 Aluminum, Caffeine Free Diet Pepsi, test market small globe, 1997, 12 oz., $5.00

Left: 139 Aluminum, Pepsi Free, c.1980s, 12 oz., $5.00
Right: 140 Aluminum, Diet Pepsi Free, c.1980s, 12 oz., $5.00

Left: 146 Aluminum, Wild Cherry Pepsi, c.1980s, 12 oz., $5.00
Right: 147 Aluminum, Wild Cherry Diet Pepsi, c.1980s, 12 oz., $5.00

From left: 141 Aluminum, Raging Razzberry Pepsi, test item, 1991, 12 oz., $10.00
142 Aluminum, Strawberry Burst Pepsi, test item, 1991, 12 oz., $10.00
143 Aluminum, Tropical Chill Diet Pepsi, test item, 1991, 12 oz., $10.00

Left: 144 Aluminum, Pepsi A. M., test item, 1987, 12 oz., $10.00
Right: 145 Aluminum, Diet Pepsi A.M., test item, 1987, 12 oz., $10.00

Left: 148 Aluminum, Crystal Pepsi, 1992, 12 oz., $5.00
Right: 149 Aluminum, Diet Crystal Pepsi, 1992, 12 oz., $5.00

150 Aluminum, Crystal...from the makers of Pepsi, test item, 1994, 12 oz., $5.00

Left: 151 Aluminum, Pepsi X L, test item, 1995, 12 oz., $5.00
Right: 152 Aluminum, Pepsi X L, test item, 1995, 16 oz., $10.00

From left: 155 Steel, Mountain Dew with hillbilly, c.1960s, 12 oz., $25.00
156 Steel, Mountain Dew, c.1970s, 12 oz., $5.00
157 Steel, Mountain Dew, c.1970s, 12 oz., $5.00

Left: 153 Aluminum, Pepsi Kona Coffee Cola, test item, 1996, 12 oz., $10.00
Right: 154 Aluminum, Diet Pepsi Kona Coffee Cola, test item, 1996, 12 oz., $10.00

Left: 158 Aluminum, Diet Mountain Dew, 1987, 12 oz., $5.00
Right: 159 Aluminum, Diet Mountain Dew, c.1990s, 12 oz., $5.00

From left: 168 Aluminum, Aspen, test item, 1978, 12 oz., $15.00
169 Aluminum, Mountain Dew Red, test item, 1988, 12 oz., $20.00
170 Aluminum, Jake's Diet Cola, test item, 1987, 12 oz., $15.00

From left: 160 Aluminum, Caffeine Free Mountain Dew, c.1990s, 12 oz., $3.00
161 Aluminum, Caffeine Free Diet Mountain Dew, c.1990s, 12 oz., $3.00
162 Aluminum, Caffeine Free Mountain Dew, c.1990s, 12 oz., $3.00
163 Aluminum, Caffeine Free Diet Mountain Dew, c.1990s, 12 oz., $3.00

164 Steel, Sugar Free Mountain Dew, c.1970s, 12 oz., $20.00

From left: 165 Aluminum, Mountain Dew Sport, 1989, 12 oz., $10.00
166 Aluminum, Mountain Dew Sport, 1989, 12 oz., $10.00
167 Aluminum, Mountain Dew Diet Sport, 1989, 12 oz., $10.00

171- 174 Four aluminum cans, H2 Oh! (shown in various flavors), test items, 1989, 12 oz., $5.00 each Aluminum, H2 Oh!, test item, 1989, 12 oz., $5.00

175 Steel, Patio Diet Cola, 1963, 12 oz., $35.00

176-178 Three steel cans, Patio (shown in various flavors), c.1960s, 12 oz., $20.00 each

179-181 Three steel cans, Patio (shown in various flavors), c.1960s, 12 oz., $20.00 each (*left & center*), $10.00 (*right*)

182-184 Three aluminum cans, Slice (shown in various flavors), c.1980s, 12 oz., $5.00 each

185-187 Three aluminum cans, Slice (shown in various flavors), c.1980s, 12 oz., $5.00 each

188-190 Three aluminum cans, Slice (shown in various flavors), c.1980s, 12 oz., $5.00 each

191-193 Three aluminum cans, Slice (shown in various flavors), c.1980s, 12 oz., $5.00 each

Left: 194 Steel, Teem, c.1960s, 12 oz., $20.00
Right: 195 Steel, Teem, c.1970s, 12 oz., $15.00

196 Steel, textured display can, c.1960s, 12 oz., RARE

197 Steel, swirl test can, front and back, c.1960s, 12 oz., RARE

198-199 Steel, swirl test can (shown in silver and white), c.1960s, 12 oz., RARE

200 Steel, straight sided test can, front and back, c.1960s, 12 oz., RARE

201 Steel, straight sided test can, front and back, c.1960s, 12 oz., RARE

Carriers

202 Cardboard, Pepsi Light, c.1970s, $5.00

203 Cardboard, Pepsi Light, c.1970s, $5.00

204 Cardboard, Pepsi, c.1970s, $5.00

205 Cardboard, Apple Slice, c.1980s, $5.00

206 Cardboard, Teem, c.1960s, $10.00

207 Cardboard, Sugar Free Teem, 1964, RARE

208 Cardboard, Diet Pepsi, c.1970s, $5.00

209 Cardboard, Diet Pepsi, c.1970s, $5.00

210 Cardboard, Pepsi Free, c.1980s, $5.00

211 Cardboard, Devil Shake, c.1960s, $30.00

212 Cardboard, Aspen, test item, 1978, $15.00

213 Cardboard, Patio Diet Cola, 1963, $15.00

214 Cardboard, Pepsi, Canadian, c.1950s, $15.00

215 Cardboard, Pepsi, Canadian, c.1950s, $15.00

217 Cardboard, without bottles, Pepsi, c.1990s, $5.00

218 Cardboard, Petty Longnecks, 1992, $10.00

216 Cardboard, Pepsi, c.1940s, $75.00

219 Cardboard, without bottles, Pepsi, 1998, $5.00

220 Cardboard, without bottles, Caffeine Free Pepsi, c.1980s, $5.00

221 Cardboard, Pepsi, c.1980s, with bottles, $25.00

222 Cardboard, Pepsi, c.1980s, $10.00

223 Cardboard, Albuquerque Balloon Fiesta, 1995, $5.00

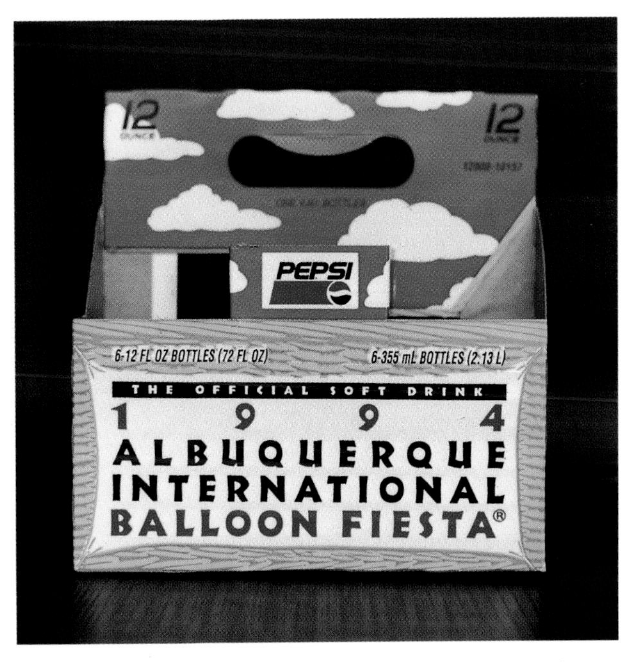

224 Cardboard, Albuquerque Balloon Fiesta, 1994, $5.00

225 Cardboard, H2 Oh!, test item, 1989, $10.00

226 Cardboard, Pepsi, c.1980s, $5.00

227 Cardboard, Pepsi, c.1980s, $5.00

228 Cardboard, Diet Pepsi, c.1980s, $5.00

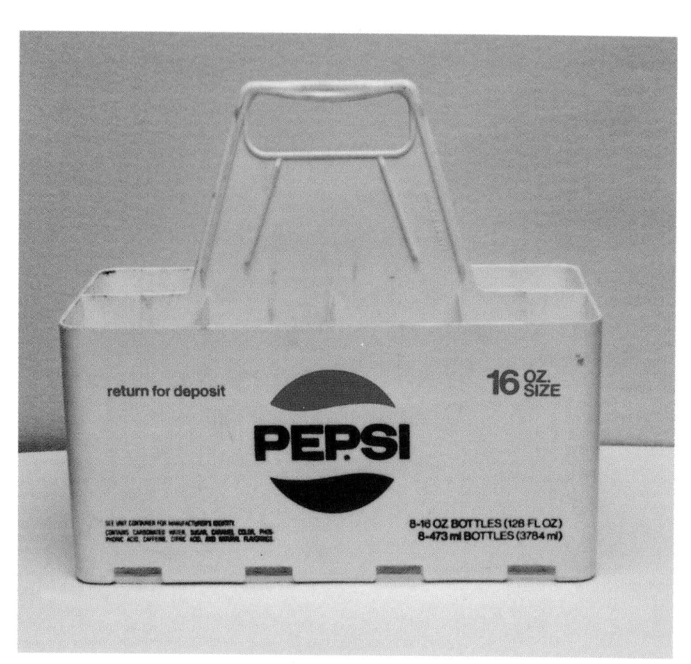

231 Plastic, Pepsi, c.1970s, $15.00

230 Plastic, Pepsi, c.1970s, $10.00

229 Plastic, Pepsi, c.1960s, $15.00

233 Heavy cardboard with plastic rim, Pepsi, c.1980s, $20.00

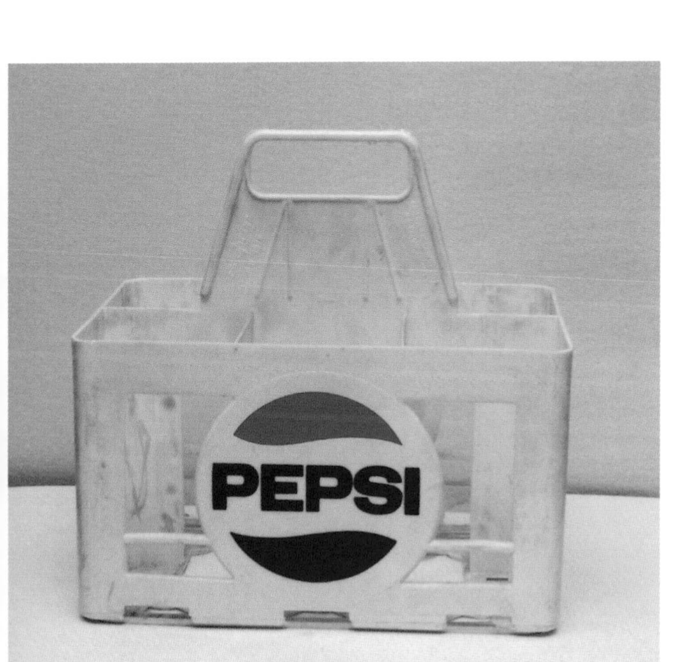

232 Plastic, Pepsi, c.1970s, $15.00

234 Wood, Pepsi, with bottles, c.1970s, $85.00

235 Metal, Pepsi, c.1950s, $50.00

236 Plastic, Pepsi, c.1970s, $15.00

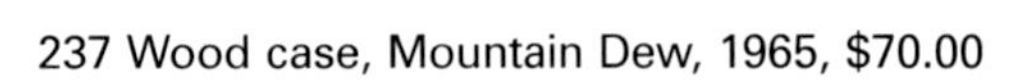

237 Wood case, Mountain Dew, 1965, $70.00

238 Wood case, Pepsi, Newberry, Michigan, c.1950s, $50.00

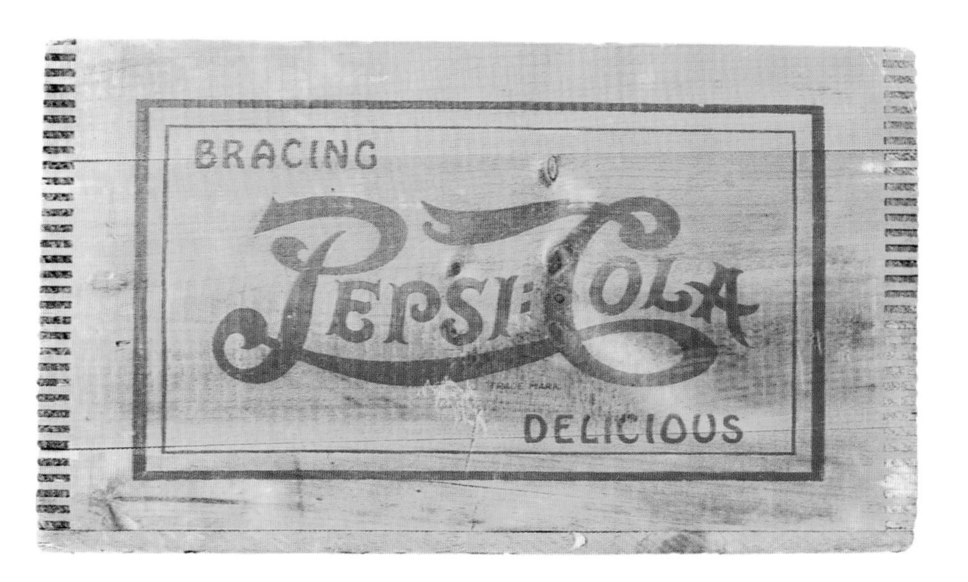

239 Wood case, Pepsi, c.1930s, $100.00

240 Cardboard shells, Pepsi, c.1970s, $5.00

241 Cardboard case, Pepsi, c.1960s, $15.00

242 Cardboard case, Pepsi, c.1960s, $20.00

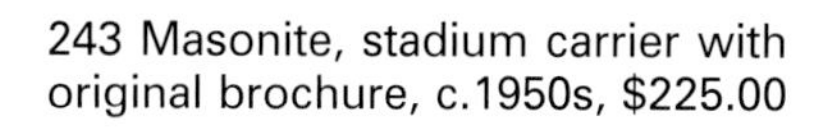

243 Masonite, stadium carrier with original brochure, c.1950s, $225.00

244 Metal, stadium carrier, c.1950s, $200.00

Clocks

47 Electric with calendar pages, c.1960s, $150.00 (also came lighted)

246 Electric, lighted, animated, Teem, c.1960s, $450.00

248 Electric, lighted with calendar pages, c.1960s, $125.00

245 Electric, kitchen type, c.1940s, $75.00

249 Plastic, electric, c.1970s, $30.00

254 Wind-up, c.1980s, $20.00

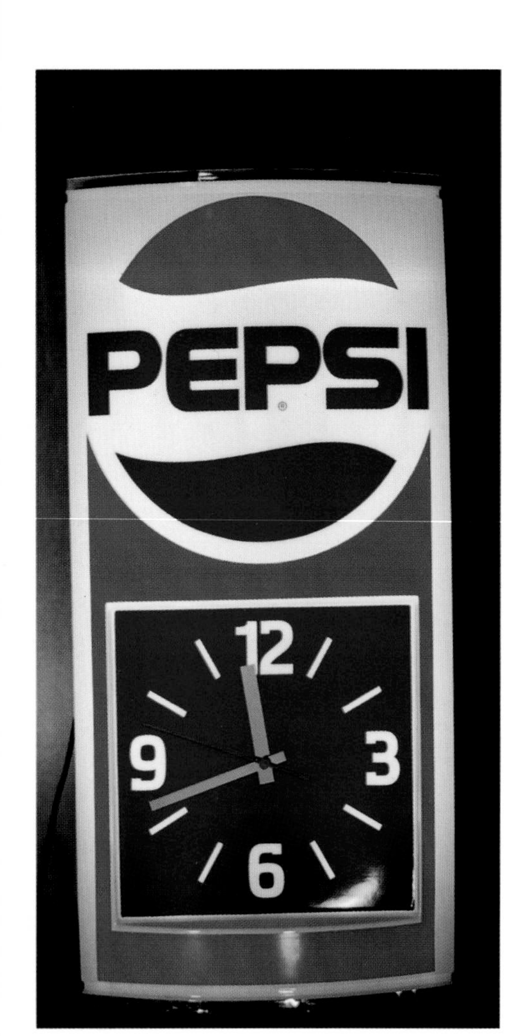

252 Electric, lighted, c.1980s, $50.00

250 Wind-up travel alarm, c.1970s, $40.00

251 Quartz with leather cover travel alarm, 1985, $30.00

253 Quartz, c.1980s, $20.00

255 Quartz, reverse on glass, $50.00

258 Quartz, plastic, c.1990s, $20.00

257 Quartz, wood, c.1980s, $50.00

256 Quartz, wood, c.1980s, $50.00

260 Quartz, plastic, c.1990s, $30.00

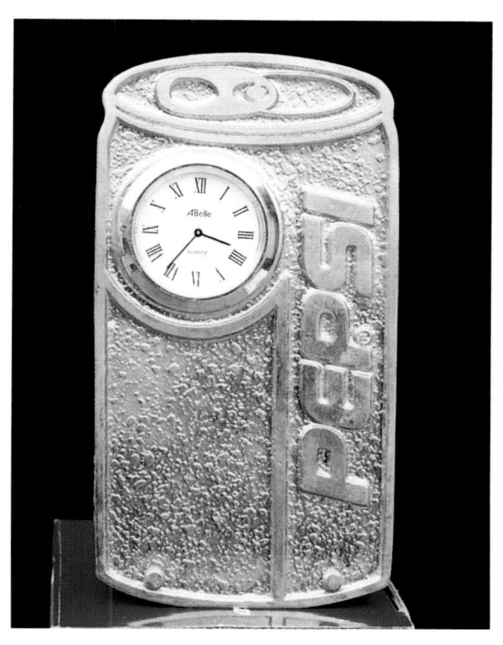

259 Quartz, embossed brass, c.1990s, $20.00

261 Quartz, glass, c.1990s, $40.00

262 Quartz, wood, Diet Pepsi, c.1990s, $75.00

264 Quartz, wood, reverse on glass, 15" x 26", c.1990s, $75.00

263 Quartz, plastic, Crystal Pepsi, c.1990s, $30.00

Clothing

265 Driver's shirt, front, c.1930s, $100.00

265 Driver's shirt, back

Above: 266 Bike racing shirt, nylon, c.1980s, $25.00

Left: 267 Jacket, Pepsi Challenger, c.1980s, $35.00

268 Varsity jacket, Crystal Pepsi, c.1990s, $100.00

269 Handkerchiefs with plastic box, c.1980s, $15.00

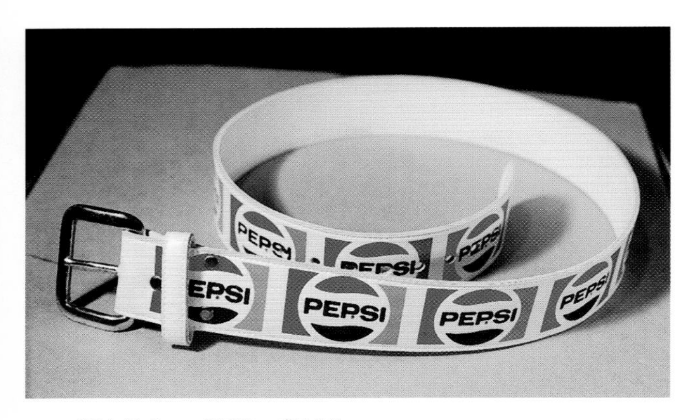

271 Belt, c.1980s, $5.00

270 Slippers, can graphics, c.1990s, $10.00

272 Baby bib, c.1990s, $5.00

273 Baby bib, c.1990s, $5.00

274 Tennis shoes with box, c.1980s, $20.00

275 Tennis shoes with box, c.1980s, $20.00

276 Felt hats, Mountain Dew, c.1960s, $10.00 each

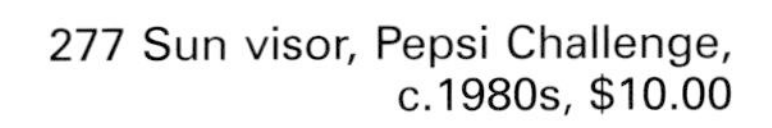

277 Sun visor, Pepsi Challenge, c.1980s, $10.00

278 Scarf, Evervess, c.1940s, RARE

Commemorative Items—100th Anniversary

279 Pewter on wood plaque, 1 of 100, 1998, $100.00

280 Brass logo medallions, framed and matted, 1998, $250.00

281 Bottle cap pins, framed and matted, 1998, $75.00

282 Canvas 6-pack bag, 9" x 4", 1998, $10.00

283 Canvas tote bag, 18" x 18", 1998, $10.00

284 Nylon tote bag, Hawaii, 1998, $25.00

285 Hat pins, eight different, 1998, $5.00 each

287 Gold charm, 14 carat, Hawaii, 3/4", 1998, $100.00

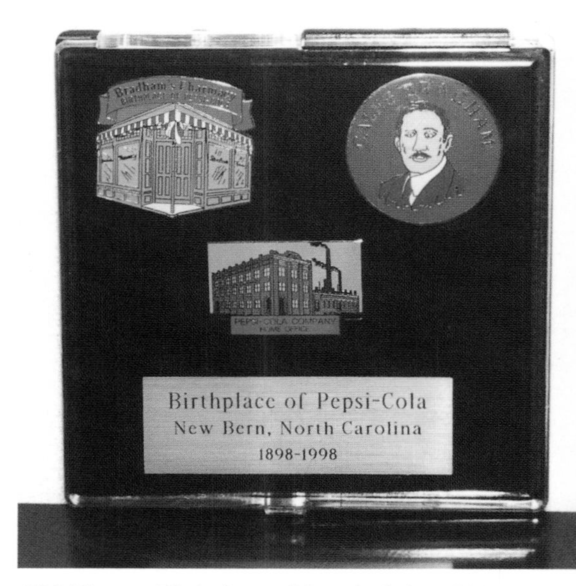

286 Pin set, Birthplace of Pepsi- Cola, 1998, $15.00

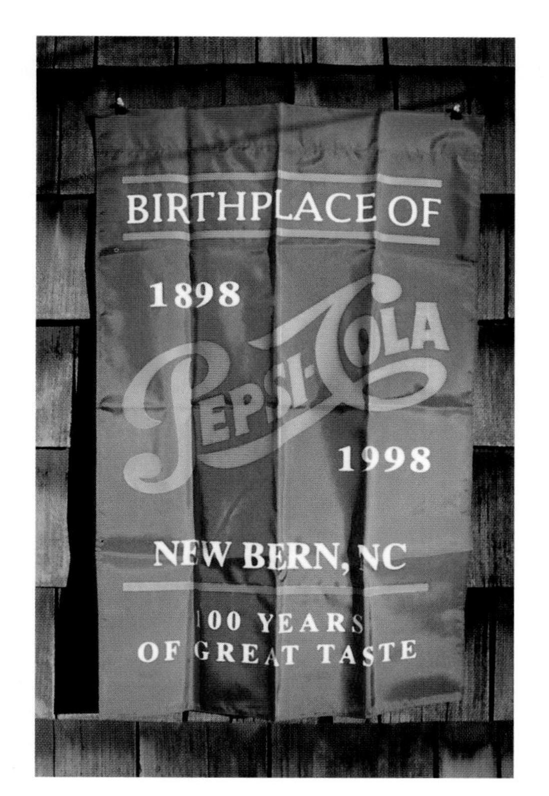

288 Banner, Birthplace of Pepsi- Cola, 1998, $50.00

289 Original street banner with certificate, 1998, $75.00

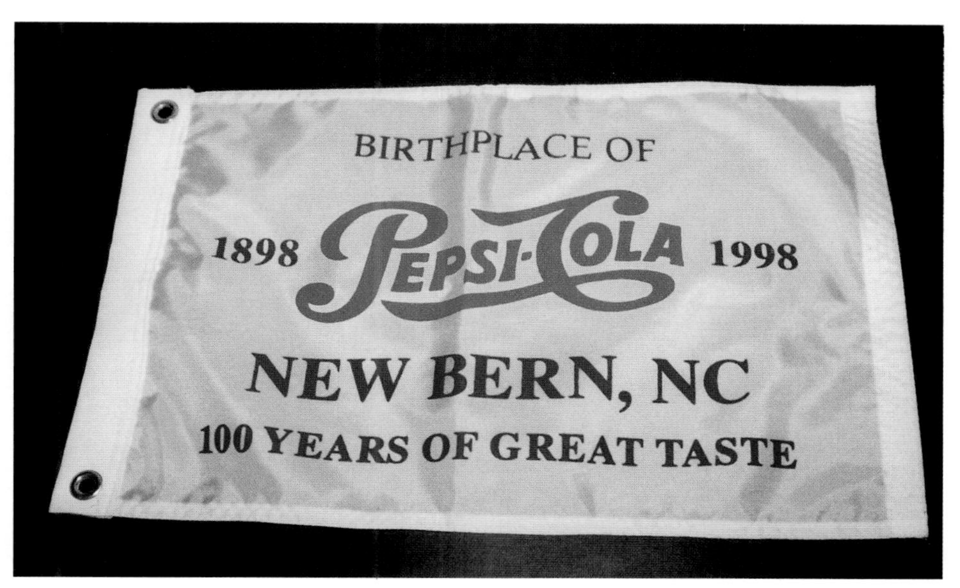

290 Boat flag, Birthplace of Pepsi- Cola, 20" x 13", 1998, $20.00

291 Boat flag, Birthplace of Pepsi- Cola, 20" x 11.5", 1998, $20.00

292 Postcard, Hawaii, 1998, $5.00

293 Large bottle, 1 of 500, front and back view, 19" tall, 1998, $65.00

294 Hat pin with card, Hawaii, 1", 1998, $5.00

295 Flyer with ticket, autographed by Charlie Daniels, 1998, $35.00

296 Can, Pepsi's 100, 12 oz., 1998, $5.00

297 Penny set, Birthplace of Pepsi- Cola, 1998, $20.00

298 Rhea egg, hand carved, one of a kind, 1998, $150.00

299 Watch fob with certificate, 1 of 100, 1998, $50.00

300 Truck bank, 1/64th scale, diecast, 1998, $35.00

301 Ceramic tankard, Pepsi and Pete, 1998, $15.00

302 Ceramic tankard, diamond logo, 1998, $15.00

303 Pillow with trucks, Hawaii, 20" x 12", 1998, $55.00

305 Windsock with centennial logo, 1998, $20.00

304 Pillow with centennial logo, Hawaii, 13" x 13", 1998, $55.00

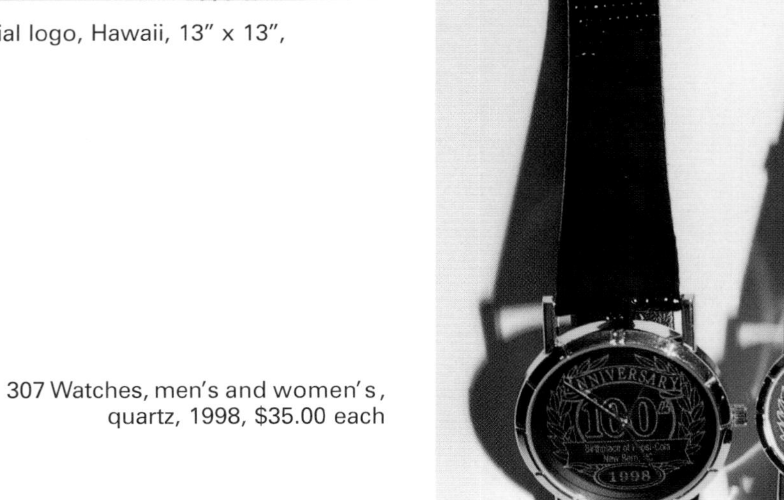

307 Watches, men's and women's, quartz, 1998, $35.00 each

306 Plastic tumbler, Birthplace of Pepsi- Cola, 12 oz., 1998, $2.00

308 Key ring, bronze, 1998, $5.00

309 Key ring, bronze, 1998, $5.00

310 Key ring, plastic, 1998, $5.00

311 Key ring, brass, front, 1998, $10.00

311 Key ring, brass, back

312 Cloth patches, 4 different, 1998, $5.00 each

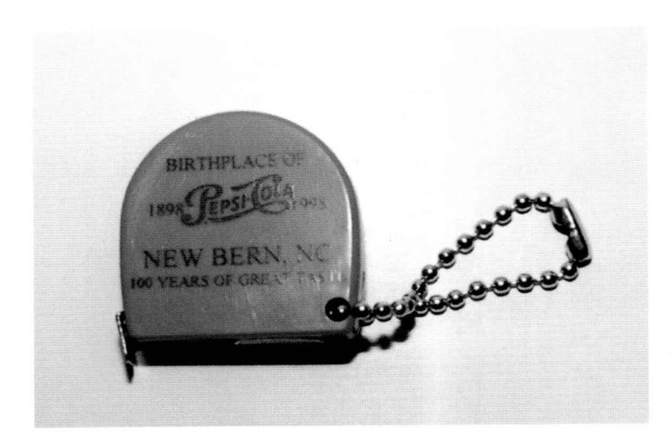

313 Tape measure, plastic, 1998, $5.00

314 Silk tie, centennial logos, Hawaii, 1998, $50.00

315 Clock, quartz, wood, reverse on glass, 100th Anniversary on face, 15" x 26", 1998, $100.00

316 Clock, quartz, wood, reverse on glass, 100 Years on face, 15" x 26", 1998, $125.00

317 Coaster set, Birthplace of Pepsi- Cola, 1998, $35.00

318 Teddy bears, 5 different, Birthplace of Pepsi- Cola, 8" tall, 1998, $20.00 each

319 Plate with box, Lennox, 10", 1998, $125.00

320 Bottle set, 4 different with carrier and shelf talker, 1998, $25.00

321 Bottle set, 8 different with wood display case, 1998, $300.00

322 First day covers, set of five, 1998, $10.00

323 First day covers, set of 15, 1998, $30.00

324 First day covers, set of 15, 1998, $30.00

325 Playing cards, 1998, $5.00

326 Serving tray, 1998, $10.00

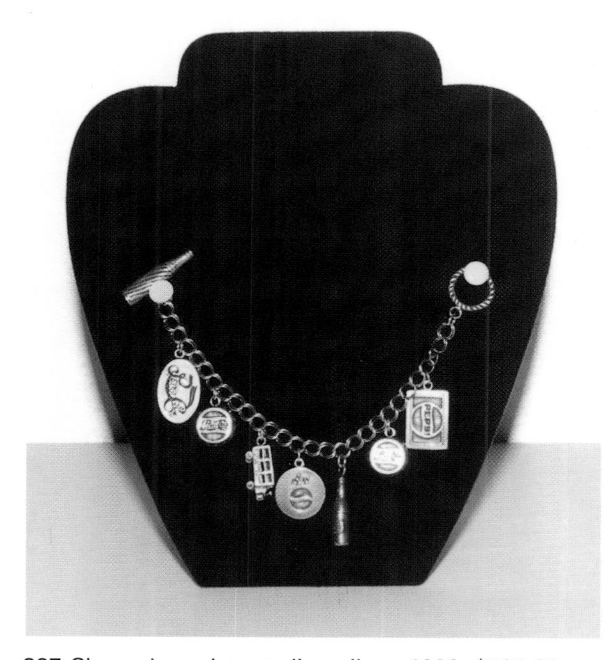

327 Charm bracelet, sterling silver, 1998, $125.00

328 Scratch pad cube, 1998, $10.00

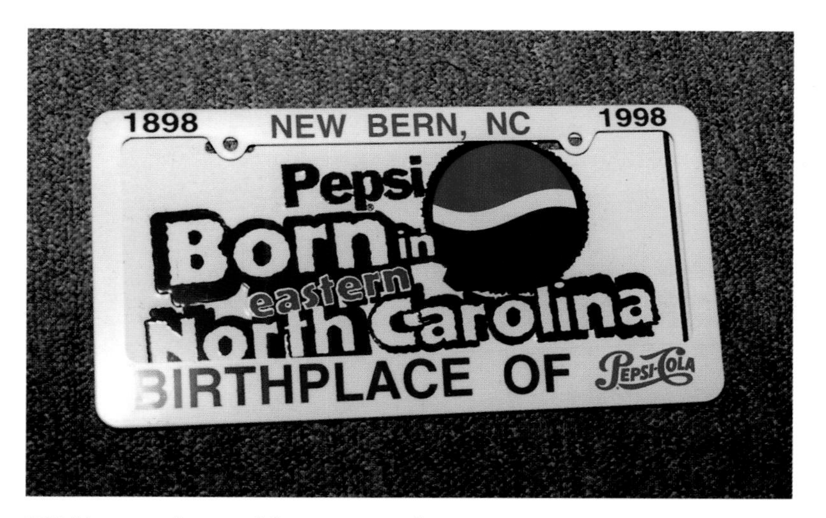

329 License plate and frame, 1998, $12.00 set

330 Bottler's agenda booklet, Pepsi's 100th in New Bern, front, 1998, $30.00

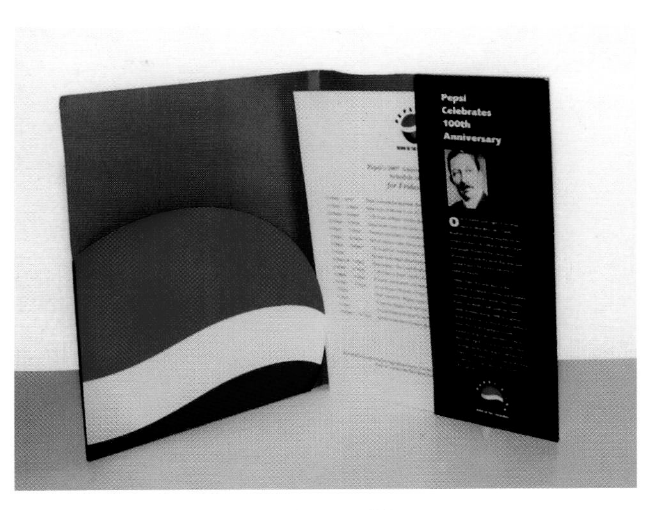

330 Bottler's agenda booklet, inside

331 Paper bag, Hawaii, 1998, $5.00

Coolers

332 Vinyl, rectangular, dark blue, c.1950s, $75.00

333 Vinyl, rectangular, "The Light Refreshment," c.1950s, $100.00

334 Vinyl, round, "Be Sociable," c.1950s, $50.00

335 Vinyl, round, "say Pepsi, please," c.1950s, $40.00

336 Vinyl, square, luggage stickers, c.1960s, $40.00

337 Vinyl, round, luggage stickers, c.1960s, $40.00

338 Vinyl, rectangular, "say Pepsi, please," c.1960s, $40.00

339 Vinyl, round, Disneyland fun, New York World's Fair, 1964-1965, $50.00

340 Vinyl, round, logos on lid, c.1960s, $30.00

342 Vinyl, round, balloon, "Feelin' Free," c.1970s, $30.00

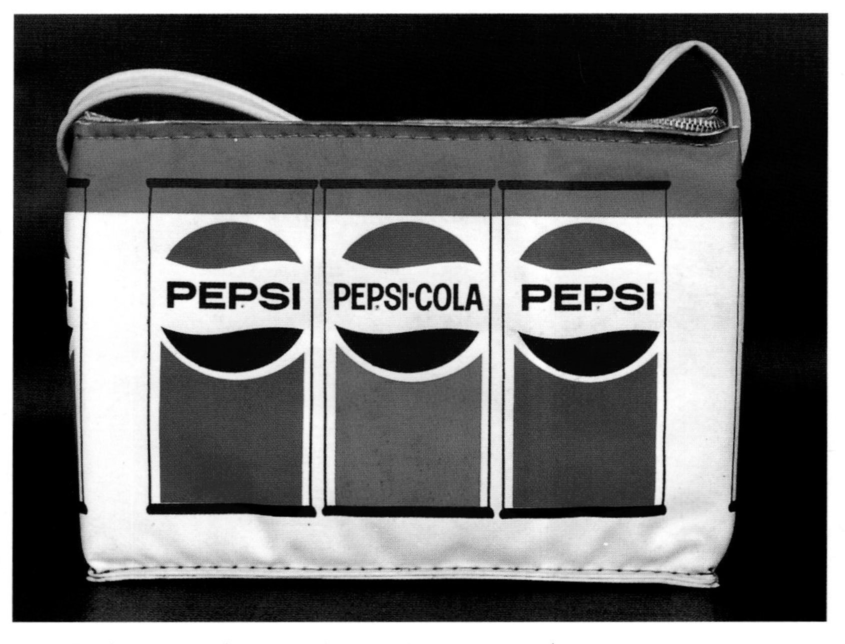

341 Vinyl, rectangular, 6-pack graphics, c.1970s, $20.00

343 Vinyl, round, bookend logo, c.1970s, $20.00

344 Vinyl, rectangular, 12-pack graphics, c.1980s, $20.00

346 Vinyl, square, speckled logo, c.1980s, $10.00

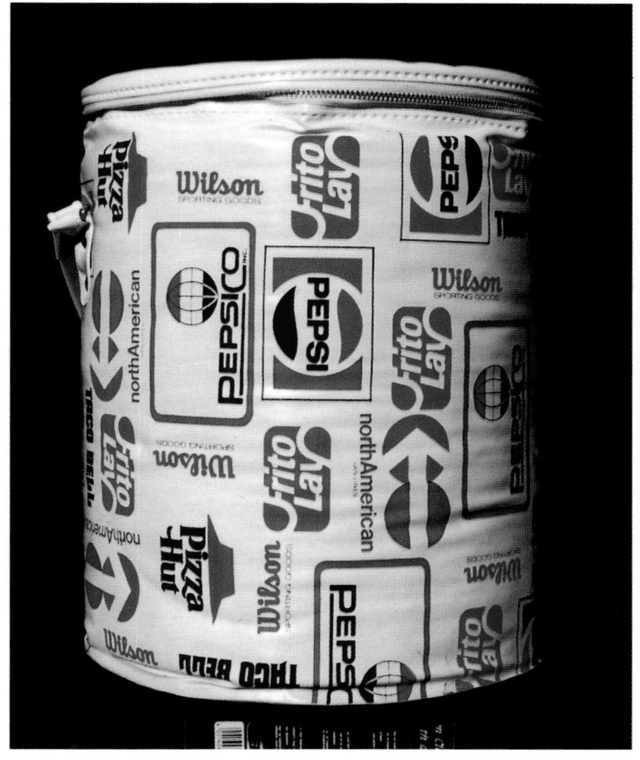

345 Vinyl, round, Pepsico, c.1980s, $20.00

347 Vinyl, square, bookend logo, c.1980s, $10.00

348 Nylon, rectangular, side opening, c.1990s, $10.00

349 Vinyl, rectangular, "Be Young, Have Fun...," c.1990s, $10.00

350 Vinyl, rectangular, can on ice graphics, c.1990s, $10.00

351 Cloth, rectangular, "Feelin' Free," c.1970s, $35.00

352 Metal with plastic insert, round, can graphics, c.1980s, $30.00

353 Plastic, round, can graphics, 8" x 13.5", c.1990s, $20.00

354 Plastic, round, can graphics, 12" x 20.5", c.1990s, $30.00

355 Plastic, round, can graphics, 8" x 13.5", 1998, $20.00

Desk Items

356 Paperweight, glass, 4" x 4", c.1900s, RARE

357 Paperweight, glass, 3", c.1950s, $45.00

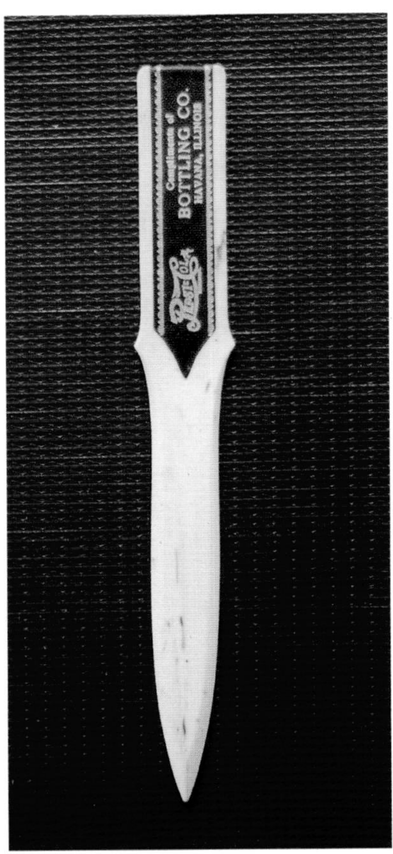
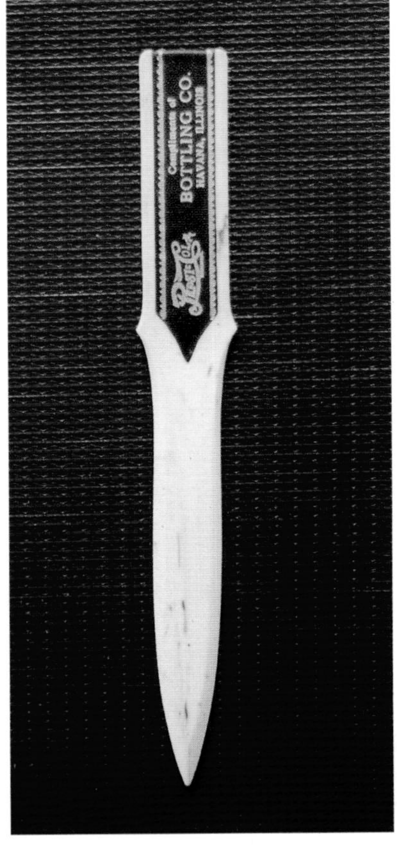

358 Letter opener, celluloid with leather, 8.5", c.1940s, $55.00

359 Pen holder, plastic, c.1960s, $40.00

361 Business card holder, metal, 4", c.1980s, $10.00

362 Business card folder, vinyl, c.1970s, $10.00

360 Letter holder, ceramic tile with wire clip, 4.5" x 4.5", 1986, $20.00

363 Tape measure, Pepsi Challenge, c.1980s, $15.00

365 Pencil sharpener, plastic, can graphics, 2.5", c.1970s, $10.00

364 Tape measure with box, bookend logo, c.1970s, $15.00

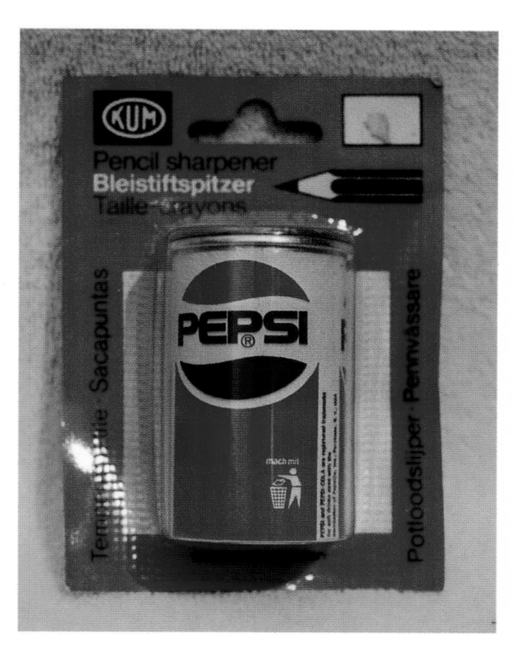

366 Pencil sharpener in package, can graphics, 2.5", c.1980s, $5.00

367 Pencil bag, cloth, 8.25" x 4.75", c.1990s, $5.00

368 Pencil bag, vinyl, 8" x 3", end opening, c.1990s, $5.00

369 Scratch pad cube, bottle graphics, c.1980s, $5.00

371 Scratch pad cube, Crystal Pepsi, 1993, $10.00

370 Scratch pad cube with wood pallet, c.1980s, $10.00

Drinking Glasses

Left: 372-373 Two sample cups, wax paper, 6 oz. and 2 oz., c.1960s, $10.00 each
Right: 374 Sample cup, wax paper, 2 oz., 1977, $5.00

375 Two sample cups, wax paper, 12 oz. and 6 oz., 1967, $15.00-20.00 each

Left: 377 Sample cup, wax paper, 12 oz., 1995, $3.00
Right: 378 Sample cup, plastic, 8 oz., 1998, $1.00

Left: 379 Sample cup, paper, 12 oz., 1993, $3.00
Right: 380 Sample cup, plastic, 6 oz., 1993, $3.00

381 Tumbler, plastic, c.1950s, $22.00

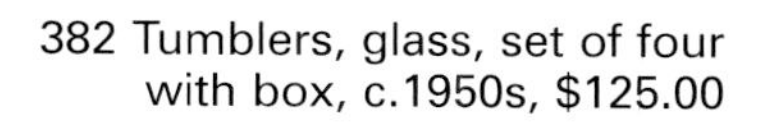

382 Tumblers, glass, set of four with box, c.1950s, $125.00

383 Tumbler, glass, Plant Opening, c.1960s, $25.00

384 Two tumbler, glasses (two sizes shown), c.1970s, $10.00 each

386 Pitcher, glass, Tiffany style, c.1970s, $15.00

Food Related

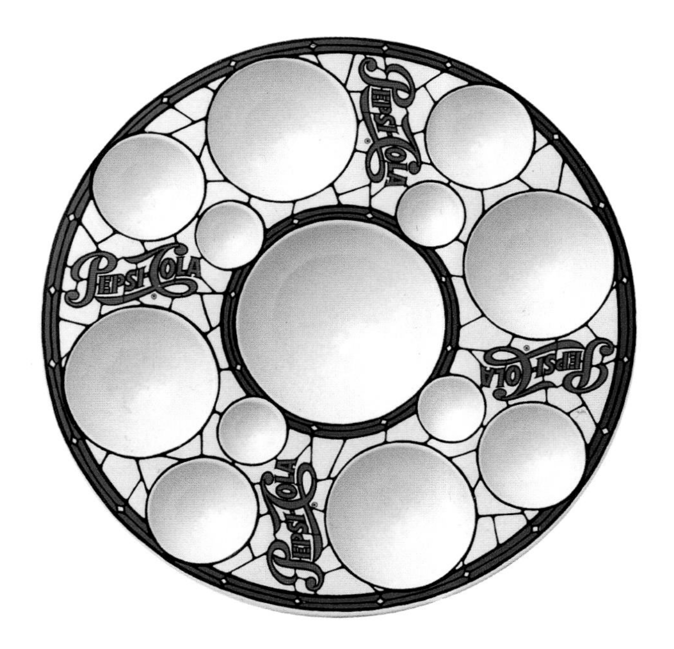

387 Chip and dip tray, plastic, Tiffany style, c.1970s, $20.00

388 Placemat, vinyl, c.1970s, $5.00

389 Grill, cast aluminum, bottle cap shape, c.1970s, $50.00

390 Drink coasters, glass, set of four with box, c.1970s, $15.00

391 Ice bucket, metal with plastic lid with box, can graphics, c.1970s, $30.00

392 Float glass set with spoons and scoop, with box, c.1970s, $40.00

393 Snack tray set with box, open, c.1970s, $40.00

393 Snack tray set, closed

394 Ice bucket and tumbler set, plastic, c.1980s, $15.00

395 Ice bucket, plastic, c.1980s, $15.00

397 Wine set with wood box, c.1980s, 40.00

396 Grill, metal, Mountain Dew can graphics, open, c.1980s, $30.00

396 Grill, metal, Mountain Dew can graphics, closed

398 Chip and dip tray, plastic, c.1990s, $5.00

399 Ice bucket with tongs, plastic, 1993, $15.00

400 Salt and pepper set, glass, Mexican, c.1950s, $75.00

401 Salt and pepper set, glass, round plastic tops, c.1960s, $30.00

402 Salt and pepper set, ceramic, Christmas style, c.1990s, $30.00

Fountain Related

403 Syrup pump, c.1940s, $65.00

404 Syrup dispenser, c.1940s, $150.00

406 Pre mix dispenser, refrigerated type, c.1980s, $200.00

405 Post mix dispenser, push button type, c.1950s, $225.00

407 CO_2 tank, c.1960s, $50.00

408 Syrup jug, glass, Teem, one gallon, c.1960s, $25.00

409 Syrup carton, cardboard, one gallon, c.1970s, $10.00

410 Fountain syrup testing kit, c.1980s, $10.00

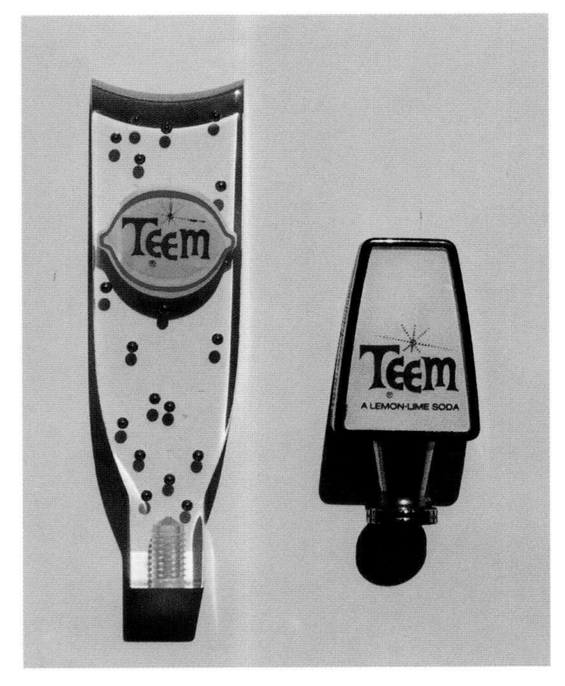

411-412 Two fountain tap knobs, Teem, c.1960s, $10.00-15.00 each

Jewelry

413 Luggage tag, metal, P G A, c.1990s, $5.00

415 Key ring, plastic, bottle shaped, c.1980s, $10.00

414 Knife and nail clipper set, c.1990s, $10.00

416 Key ring, plastic, Crystal Pepsi, 1993, $5.00

418 Key ring, plastic, bottle cap logo, Safety Award, c.1960s, $25.00

420 Money clip, pewter, c.1970s, $10.00

417 Key ring, metal, c.1980s, $10.00

419 Key ring, plastic, bookend logo, c.1970s, $20.00

421 Earrings, metal with enamel, bookend logo, c.1980s, $10.00

422 Necklace, small can, c.1980s, $5.00

424 Stick pin, metal with enamel, c.1980s, $10.00

423 Necklace with case, Skandi, 1967, $40.00

425 Medallion with link chain, metal with enamel, c.1950s, $25.00

Top left & right: 426 Participation medals, large with ribbon, c.1980s, $15.00 each
Bottom left & right: 427 Participation medals, small with chain, c.1980s, $10.00 each

Knives

428 Retractable blade, autopoint, c.1950s, 4", $25.00

429 Multiple blades with wrench, c.1960s, $30.00

430 Pocket knife, c.1970s, 2", $10.00

431 Pocket knife, c.1970s, 1.75", $10.00

433 Pocket knife, Victorinox, c.1970s, 3.5", $25.00

432 Pocket knife, c.1970s, 2", $10.00

434 Pocket knife, Diet Pepsi, c.1970s, 3.75", $25.00

435 Pocket knife, c.1980s, 4", $20.00

437 Multiple blade pocket knife, c.1980s, 2.75", $10.00

436 Round, 75th Anniversary, front, 1973, 1.5", $10.00

436 Round, 75th Anniversary, back

Lights

438 Hanging lamp, plastic, c.1970s, $150.00

439-440 Two night lights, Pepsi Light, blue and red, c.1970s, $15.00 each

Above: 441 Night light, leaded glass, c.1990s, $15.00

Left: 442 Patio Lites set with box, c.1970s, $20.00

443 Miniature flashlight with case, c.1990s, $25.00

Left: 444 Flashlight/key chain, c.1990s, $15.00
Right: 445 Pocket flashlight, "The Light Refreshment," c.1950s, $50.00

447 Wall mount lamp with two revolving cylinders, c.1950s, $650.00

446 Flashlight/ keychain, Pepsi Light, c.1970s, $15.00

Miscellaneous

448 Print block, bottle cap, c.1950s, $10.00

449 Change tray, glass, c.1940s, $100.00

450 Case printing plate, brass, c.1960s, $50.00

451 Case printing plate, brass, c.1930s, $600.00

452 Music box, Pepsi machine, c.1990s, $75.00

453 Retractable brush, c.1960s, $10.00

454 Grocery shopping basket, plastic, c.1980s, $25.00

455 Can-shaped camera with box, c.1990s, $25.00

456 Fishing reel, Chain- O- Lakes Fishing Derby, 1970, $25.00

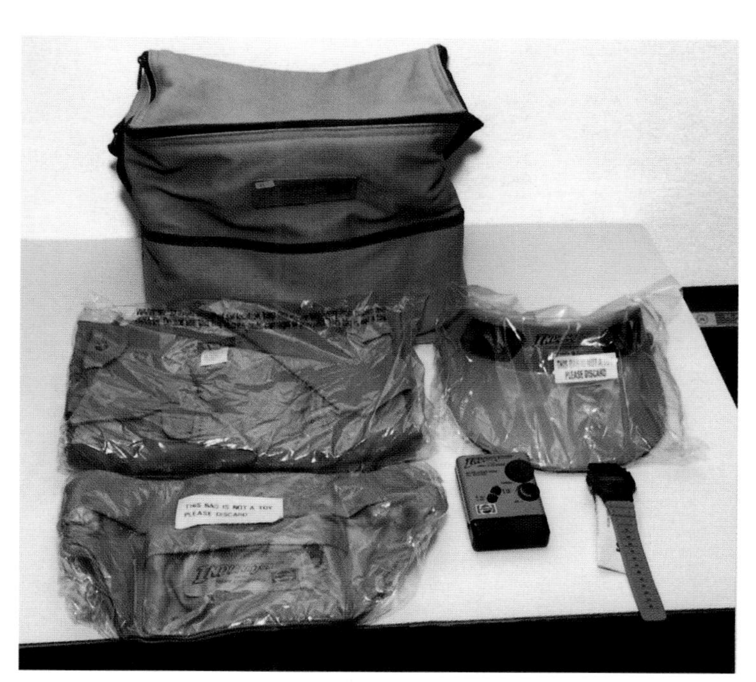

457 Indiana Jones bag (note Pepsi logo), 6 piece set, c.1980s, $60.00

458 Boxing gloves, Teem, c.1960s, $50.00

459 Sleeping bag, c.1970s, $25.00

460 Candle, bottle-shaped, Germany, c.1980s, 8", $5.00

461 Printing page, plastic, Pepsi, c.1960s, $25.00

462 Printing page, plastic, Mountain Dew, c.1960s, $30.00

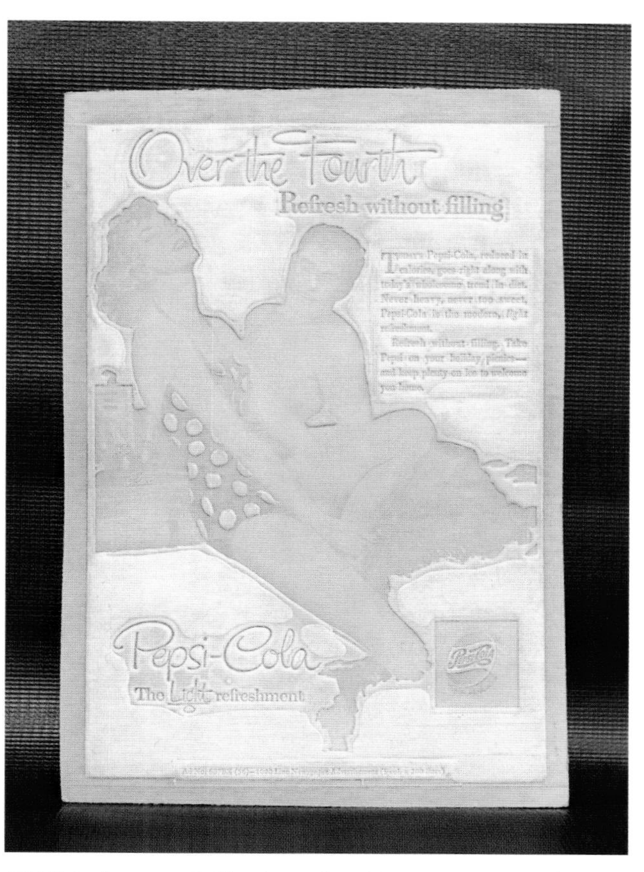

463 Printing proof, fiberboard, c.1950s, 11" x 16", $30.00

464 Compact disc carrying case, c.1990s, $10.00

465 License plate, Pepsi 400, c.1990s, $15.00

466 License plate, "Think Young," c.1960s, $25.00

467 License plate, Pepsi Challenge, c.1980s, $10.00

468 License plate, New Bern, c.1980s, $10.00

469 Telephone, can graphics, c.1980s, $70.00

470 Telephone, musical and animated, c.1990s, $100.00

471 Telephone, pick-up truck, c.1990s, $80.00

472 Trash can, c.1970s, 8" x 15", $35.00

473 Golf ball with box, c.1960s, $20.00

474 Golf ball with practice tee, c.1950s, $30.00

476 Boat Flag with patches, fishing derby, 1984, $20.00

475 Beeper, Mountain Dew, c.1990s, $40.00

477 Drink coaster, cardboard, Teem, c.1960s, $5.00

478 Go-kart, Pepsi Challenger, 3 H P gas engine, c.1980s, $700.00

Paper Items

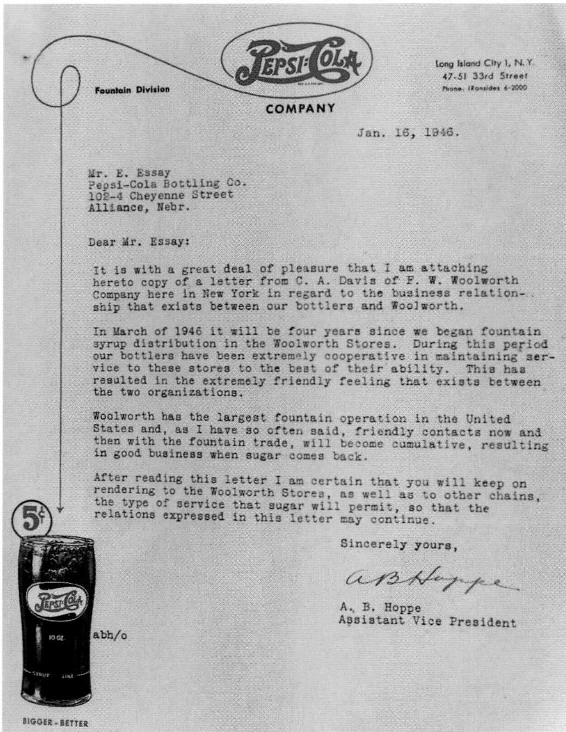

Pepsi-Cola
COMPANY

Fountain Division

Long Island City I, N. Y.
47-51 33rd Street
Phone: IRonsides 6-2000

Jan. 16, 1946.

Mr. E. Essay
Pepsi-Cola Bottling Co.
102-4 Cheyenne Street
Alliance, Nebr.

Dear Mr. Essay:

It is with a great deal of pleasure that I am attaching hereto copy of a letter from C. A. Davis of F. W. Woolworth Company here in New York in regard to the business relationship that exists between our bottlers and Woolworth.

In March of 1946 it will be four years since we began fountain syrup distribution in the Woolworth Stores. During this period our bottlers have been extremely cooperative in maintaining service to these stores to the best of their ability. This has resulted in the extremely friendly feeling that exists between the two organizations.

Woolworth has the largest fountain operation in the United States and, as I have so often said, friendly contacts now and then with the fountain trade, will become cumulative, resulting in good business when sugar comes back.

After reading this letter I am certain that you will keep on rendering to the Woolworth Stores, as well as to other chains, the type of service that sugar will permit, so that the relations expressed in this letter may continue.

Sincerely yours,

A. B. Hoppe
Assistant Vice President

abh/o

5¢
BIGGER - BETTER

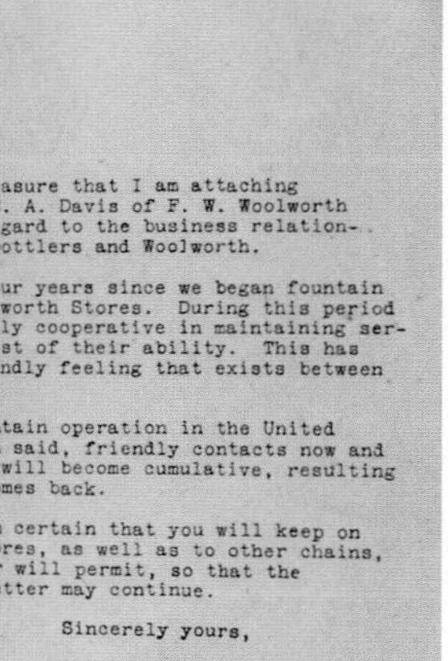

479 Letterhead, Fountain Division, 1946, $20.00

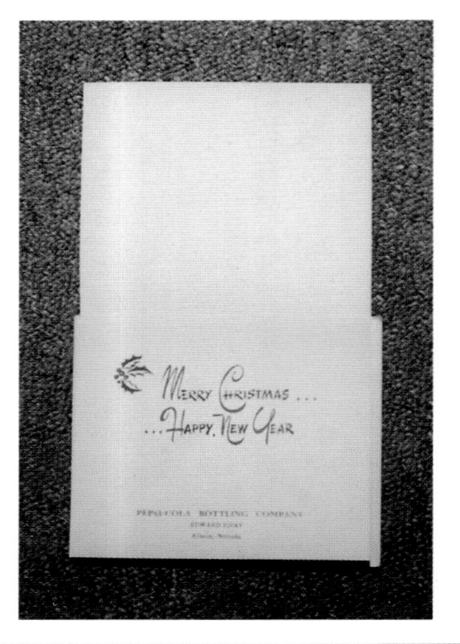

Above: 480 Christmas card, front, c.1940s, $30.00

Right: 480 Christmas card, inside

482 Booklet, advertising materials, 1941, $40.00

481 Newspaper ad, war bond, c.1940s, $15.00

483 Brochure, musical horn, front, c.1940s, $300.00

483 Brochure, musical horn, back

484 Christmas card, front, 1952, $50.00

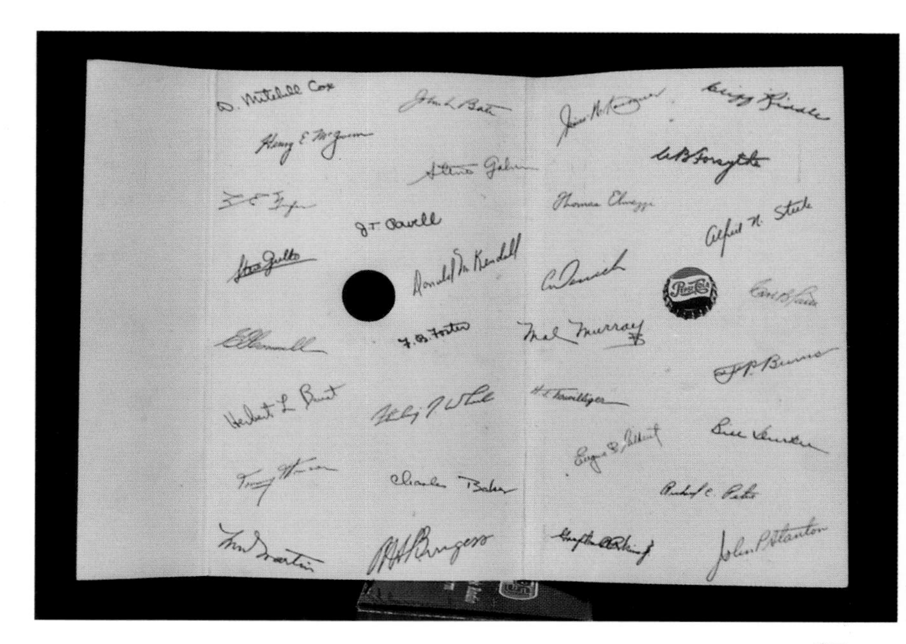

484 Christmas card, inside

485 Postcard, Boy Scouts, 6.75" x 5.5", c.1950s, $10.00

486 Advertisement, paper, c.1950s, $40.00

487 Postcard, Boy Scouts, c.1960s, $10.00

489 Game book, Mountain Dew, paper, c.1960s, RARE

488 Brochure, Mountain Dew, T V shows, outside, c.1960s, $25.00

488 Brochure, Mountain Dew, T V shows, inside

490 Brochure, Patio Diet Cola, c.1960s, $10.00

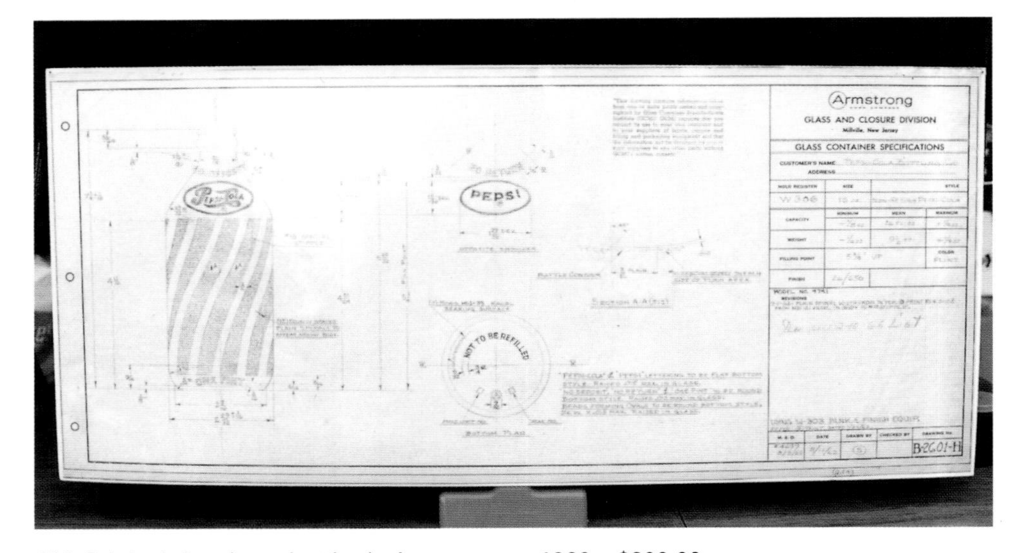

491 Original drawings, bottle design, paper, c.1960s, $200.00

492 Brochure, 6-pack vending machine, heavy paper, c.1960s, $10.00

493 Certificate, Hot Shot, paper, c.1970s, $5.00

494 Press kit, Pepsi Challenger, c.1970s, $20.00

495 Booklet, Food Service Division, c.1980s, $5.00

496 Original artwork, 1st case card, 9.25" x 18", c.1990s, $700.00

497 Original artwork, 2nd case card, 6.5" x 10.5", c.1990s, $600.00

498 Original artwork, Pepsiman, cardboard, 15" x 20", 1994, $115.00

499 Brochure, Pepsi X L, 1995, $5.00

500 Survival kit, Pepsi Stuff, c.1990s, $10.00

501 Letterhead, full package, Pepsi Stuff, c.1990s, $15.00

Pencils and Pens

502 Bullet pencil, front, c.1940s, $80.00

From top: 502 Bullet pencil, side
503 Pencil, Mountain Dew. c. 1960's, $10.00
504 Pencil, PCMI, c. 1970s, $5.00

Left: 506 Pen, Skandi, test item, 1967, $50.00
Right: 507 Pen, Tropic Surf, test item, 1967, $50.00

505 Mechanical pencil, c.1940s, $65.00

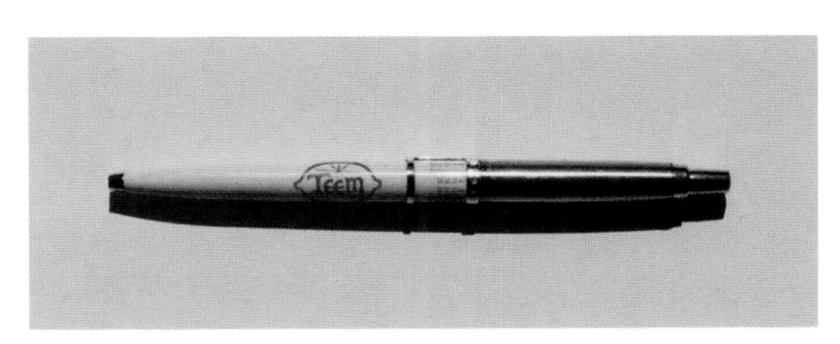

508 Pen, Teem, c.1960s, $10.00

Pinback Buttons, Patches, and Hatpins

Left: 509 Patch, Hot Shot Finalist, 3.25" x 3.75", 1979, $5.00
Right: 510 Patch, Hot Shot Winner, 3" x 3.25", c.1970s, $5.00

511 Back patch, Hot Shot Chicago, 6" x 10", c.1970s, $20.00

512 Pocket badge, Hot Shot Coordinator, c.1970s, $20.00

513 Patch, Jake's Diet Cola, test item, 1987, $10.00

From top: 514 Patch, Pepsi Light, c.1970s, $5.00
515-516 Two patches, Pepsi Light, c.1970s, $5.00
517 Back patch, Pepsi Light, c.1970s, $15.00

518 Pinback, Pepsi Challenge, 4", c.1980s, $5.00

519 Pinback, "Hey, Hotshot...," 4", c.1970s, $5.00

523 Pinback, Aspen, test item, 4", 1978, $10.00

Left: 520 Pinback, Diet Pepsi, Girl Watched Girl, 3", c.1960s, $10.00
Right: 521 Tin button, Diet Pepsi, Girl Watched Girl, 1.5", c.1960s, $5.00

524 Pinback, Jake's Diet Cola, test item, 3", 1987, $10.00

525 Pinback, Mountain Dew, "Thirsty?", 3", c.1960s, $20.00

522 Pinback, Pepsi Light, "Try New...," 3", c.1970s, $5.00

Plates

526 Pepsi-Cola girl, 7", 1984, $50.00

527 Newspaper ad, 7", 1984, $50.00

528 Logos, 7", 1986, $25.00

529 Norman Rockwell Santa, 8.25", 1989, $55.00

Point-of-Purchase (P.O.P.)

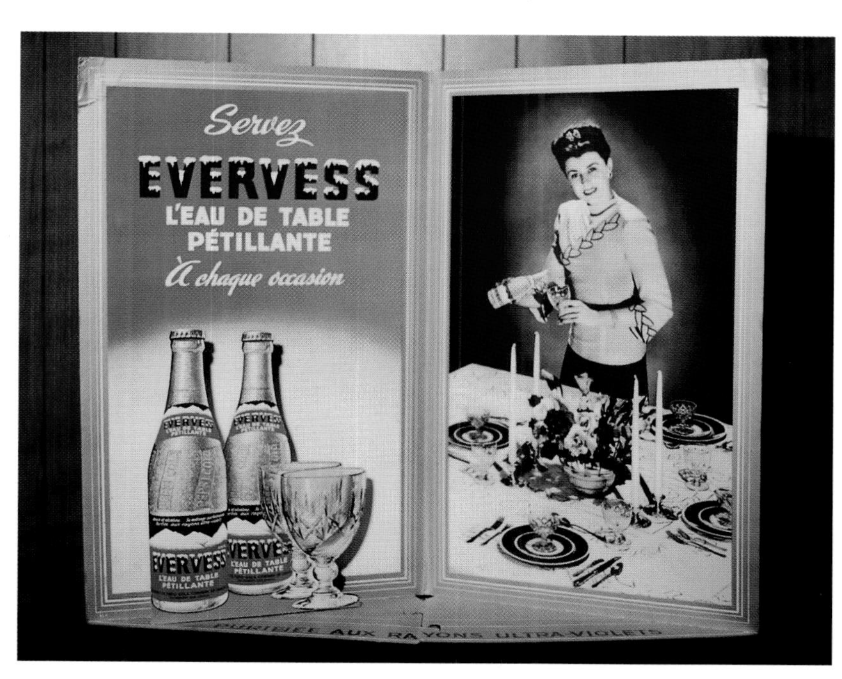

530 Evervess, 3-dimensional, cardboard, c.1940s, $175.00

532 Mountain Dew hillbilly, animated, 17" x 41", c.1960s, $600.00

531 Self-inflating snowman, lighted base, 25" x 54", c.1960s, $1,000.00

531 Self-inflating snowman, different view

533 Pouring can, Pepsi and Diet Pepsi, 12.5" x 21", c.1990s, $40.00

534 Lighted cup, Pepsi Freeze, c.1990s, $50.00

536 Revolving ball, Pepsi Stuff, 1996, $20.00

535 Rotating bottle, 1998, $10.00

539 Paper sign, Pepsi Stuff, 1996, $5.00

537 Laminated shelf talker, Pepsi Stuff, 1997, $5.00

538 Swinging street talker, Pepsi Stuff, 1996, $85.00

542 Can bank display with can, c.1960s, $150.00

544 Cardboard stand-up, A. J. Foyt, autographed, c.1990s, $25.00

541 Pepsi Challenge screen, masonite, c.1980s, $20.00

543 Cardboard sign, Pepsi Freeze, 30" x 30", c.1990s, $5.00

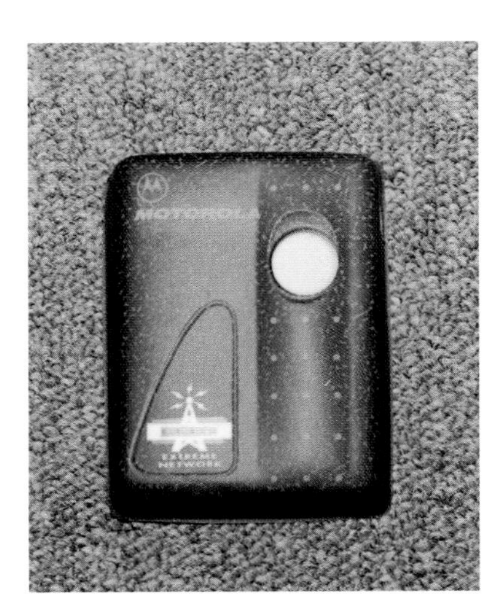

540 Cooler door talker, beeper, 3.5" x 4.5", c.1990s, $35.00

545 Cardboard stand-up, Darth Vader, 36" x 84", c.1990s, $50.00

546 Cardboard stand-up, R2 D2, 24" x 36", c.1990s, $50.00

548 Cardboard stand-up, Chewbacca, 36" x 84", c.1990s, $50.00

547 Cardboard stand-up, C3 P O, 29" x 72", c.1990s, $50.00

549 Cardboard stand-up, Cindy Crawford, 18" x 72", c.1990s, $25.00

550 Cardboard stand-up, Diet Patio Cola, Debbie Drake, 1963, $250.00

Radios

551 Battery, AM/FM, c.1980s, $40.00

552 Battery, AM/FM, c.1980s, $40.00

553 Electric, AM/FM, c.1990s, $50.00

From left: 554 Battery, Mountain Dew Can, c.1980s, $20.00
555 Battery, Mug Root Beer can, c.1980s, $20.00
556 Battery, Pepsi Free can, c.1980s, $20.00
557 Battery, Sugar Free Pepsi Free can, c. 1980s, $20.00

Santas

558 With Cane, holding can, Clothtique, 10.25", c.1990s, $80.00

559 With sack, holding bottle, Clothtique, 10.25", c.1990s, $80.00

560 Thumbs up, holding bottle, Clothtique, 7.5", c.1990s, $50.00

561 Norman Rockwell pose, holding bottle, Clothtique, c.1990s, $80.00

564 On ladder, with base, Teem, c.1960s, $75.00

562 Sitting, holding bottle and list, with Elf on 6- Pack, Clothtique, c.1990s, $100.00 for set

563 On sled, holding bottle, Clothtique, c.1990s, $80.00

Signs

566 Cardboard, corrugated, "Cool Off...," c.1940s, $75.00

565 Cardboard, girl with bottle, c.1940s, $900.00

567 Cardboard, corrugated, "Hits the Spot," c.1950s, $75.00

568 Cardboard, newspaper ad proof, 1955, $50.00

570 Cardboard, embossed, c.1960s, $75.00

569 Cardboard, Steve Allen, c.1960s, $35.00

571 Cardboard, Pepsicle, test item, c.1980s, $50.00

572 Cardboard, outfielder, c.1990s, $75.00

575 Composite, "The Spirit of '76," c.1960s, $25.00

574 Celluloid, Teem, c.1960s, $85.00

573 Celluloid, Teem, c.1960s, $85.00

576 Composite, die-cut bottle cap, c.1950s, $50.00

577 Foil, Diet Pepsi, "...and Diet too!", c.1960s, $25.00

578 Glass, bottle cap, c.1960s, $75.00

579 Glass, mirror, Texas, c.1980s, $30.00

580 Steel, single dot script, bottling plant letters, c.1950s, $750.00

581 Tin, bottle, Canadian, c.1940s, $400.00

583 Steel, flange-type, bottle cap, c.1950s, $350.00

582 Tin, Pepsi Challenge, double-sided, 60" x 37", c.1980s, $100.00

584 Tin, bottle, Canadian, c.1950s, $250.00

587 Lighted, plastic, "Large Pepsi, please!", c.1950s, $200.00

Above: 585 Tin, chalkboard, Canadian, c.1960s, $50.00

Left: 586 Lighted, glass, bottle, c.1930s, RARE

588 Lighted, neon, single dot script, c.1980s, $250.00

591 Lighted, neon, cup with ice, c.1980s, $250.00

589 Lighted, neon, bookend logo, c.1980s, $250.00

592 Lighted, neon, cup with bubbles, c.1980s, $250.00

590 Lighted, neon, globe logo, c.1990s, $200.00

593 Sticker, Mountain Dew, push/pull, c.1960s, $40.00

594 Sticker, Teem, push/pull, c.1960s, $15.00

595 Paper, "To Our Customers," 11" x 8.5", c.1940s, $50.00

598 Plastic, Pepsi Stuff, dissolve (or sliding double image) sign, 22", c.1990s, $30.00

597 Paper, framed, "Vacation North Carolina," 28" x 13", c.1960s, $35.00

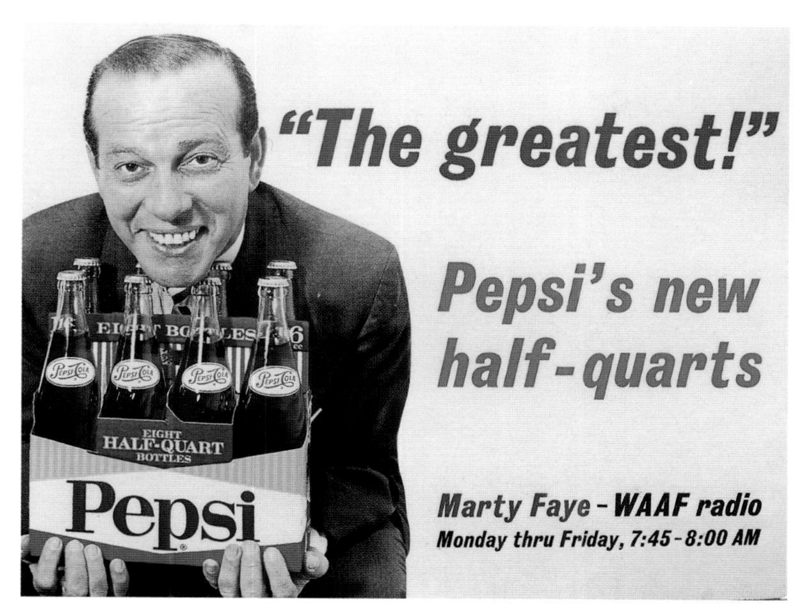

596 Paper, Half-quarts, 28" x 22", c.1960s, $30.00

599 Plastic, embossed, Teem, c.1960s, $50.00

600 Plastic, embossed, Pepsi and Diet Pepsi, c.1960s, $25.00

601 Plastic, embossed, Mountain Dew, c.1960s, $150.00

Smoking Related

604 Ashtray, glass, c.1950s, $20.00

606 Ashtray, ceramic, "Come Alive," c.1960s, $20.00

603 Ashtray, glass, c.1960s, $20.00

602 Ashtray, plastic, c.1950s, $25.00

605 Ashtray, ceramic, La Crosse, c.1960s, $30.00

607 Ashtray, glass, c.1960s, $15.00

609 Lighter, with chain, 1" x 1.5", c.1970s, $10.00

608 Ashtray, plastic, c.1960s, $45.00

613 Lighter, "Compliments of Memphis, Mo.," 2" x 1.5", c.1950s, $50.00

610 Lighter, butane, .75" x 2.75", c.1970s, $20.00

611 Lighter, butane, Diet Pepsi, .75" x 2.75", c.1980s, $20.00

612 Lighter, butane, .75" x 2.75", c.1980s, $20.00

614 Lighter, bookend logo, 1.5" x 2", c.1970s, $30.00

615 Lighter, 1.5" x 2.25", c.1960s, $40.00

616 Lighter, butane, 1" x 2.75", c.1970s, $30.00

617 Lighter, 1" x2", c.1970s, $25.00

618 Can lighter, Tropic Surf, test item, 1967, $100.00

619 Lighter, Tropic Surf, test item, 1967, $75.00

620 Lighter with box, Teem, c.1960s, $50.00

Thermometers

621 Metal, c.1950s, $150.00

623 Metal, Arabic, c.1960s, $75.00

624 Plastic, Diet Pepsi, 4.25" x 12", 1998, $5.00

622 Metal, Mexican, c.1960s, $75.00

Toys

625 Checkers table, plastic, Tiffany style, 16" x 16", c.1970s, $35.00

626 Pepsi- Swirl, with stick, box, and directions, c.1950s, $100.00

627 Kite, paper, c.1960s, $15.00

628 Dancing can with box, Diet Pepsi, battery operated, 1991, $10.00

629 Bottle bowling set, c.1960s, $55.00

630 Frisbee, Rose Bowl, 1975, $5.00

633 Bag of marbles, c.1950s, $25.00

632 Plastic model kit, c.1980s, $15.00

631 Frisbee, Pepsi Challenge, c.1980s, $5.00

634 Shooting game, c.1970s, $600.00

636 Plastic jug, Mountain Dew, front and back, c.1960s, $75.00

635 Skipper doll, 1989, $75.00

637 Puzzle, 1000 piece, c.1990s, $15.00

638 Puzzle can in package, c.1980s, $10.00

640 Puzzle, plastic can, c.1990s, $5.00

639 Puzzle, Pop-art, c.1960s, $15.00

641 Walkie Talkie set with box, c.1990s, $20.00

642 Walkie Talkie set with box, c.1990s, $20.00

644 Plastic model kit, Mountain Dew, c.1980s, $20.00

643 Doll, Mountain Dew hillbilly, 23", c.1960s, $125.00

645 Plastic model kit, Pepsi Challenger, c.1980s, $20.00

646 Plastic model kit, Darrell Waltrip, c.1990s, $20.00

Toy Trains

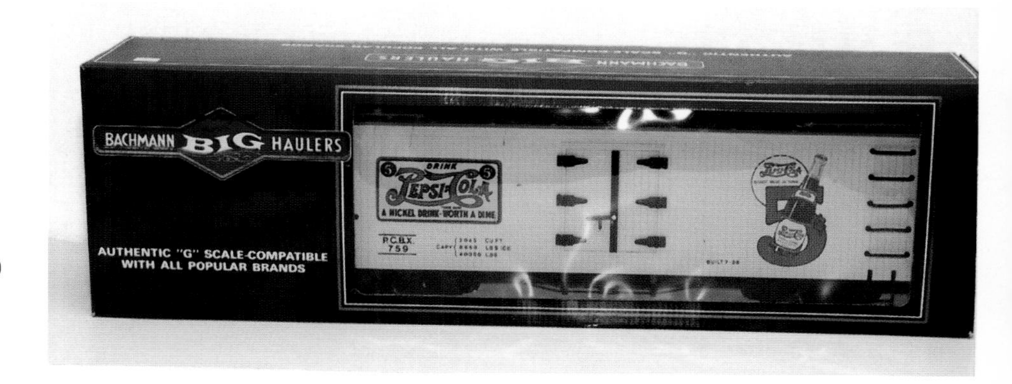

647 Box car, G scale, c.1980s, $85.00

648 Box car, 027 scale, 1982, $80.00

649 Trolley car, 027 scale, 1993, $200.00

650 Plastic model, H O scale, c.1950s, $25.00

651 Train set, 027 scale, c.1990s, $200.00

652 Train set, 027 scale, c.1990s, $200.00

653 Box car, 027 scale, c.1990s, $50.00

654 Box car, 027 scale, c.1990s, $50.00

655 Box car, 027 scale, c.1990s, $50.00

656 Box car, 027 scale, c.1990s, $50.00

657 Box car, 027 scale, c.1990s, $50.00

658 Box car, H O scale, c.1970s, $45.00

Toy Trucks

659 Bus, metal with box, c.1960s, $300.00

660 Truck, plastic, 3.5" x 1.5", Germany, c.1990s, $10.00

661 Bus, plastic in package, c.1970s, $100.00

662 Truck in package, c.1990s, $30.00

663 Truck in package, Diet Pepsi, sound machine, c.1990s, $175.00

664 Truck in package, Diet Pepsi, Tanker, c.1990s, $175.00

665 Truck in package, Sound Machine, c.1990s, $75.00

666 Tow truck in package, c.1990s, $50.00

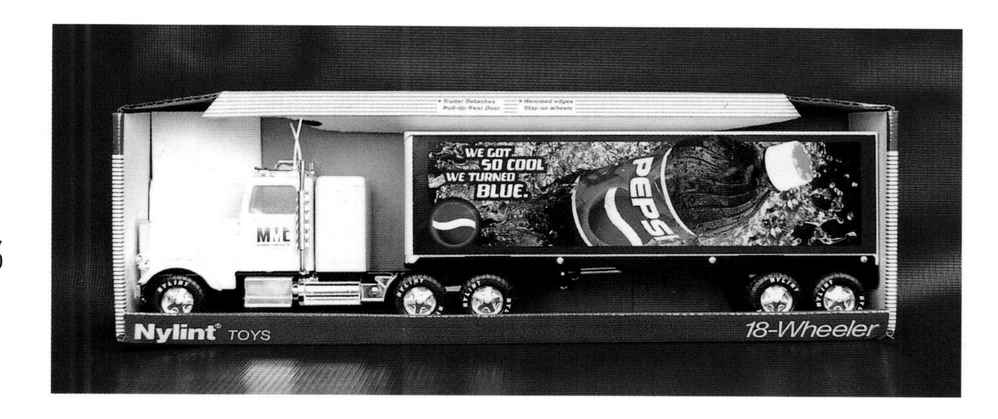

667 Truck in package, 18- Wheeler, 1998, $50.00

Trays

669 Metal, Mexican, 14", c.1950s, $75.00

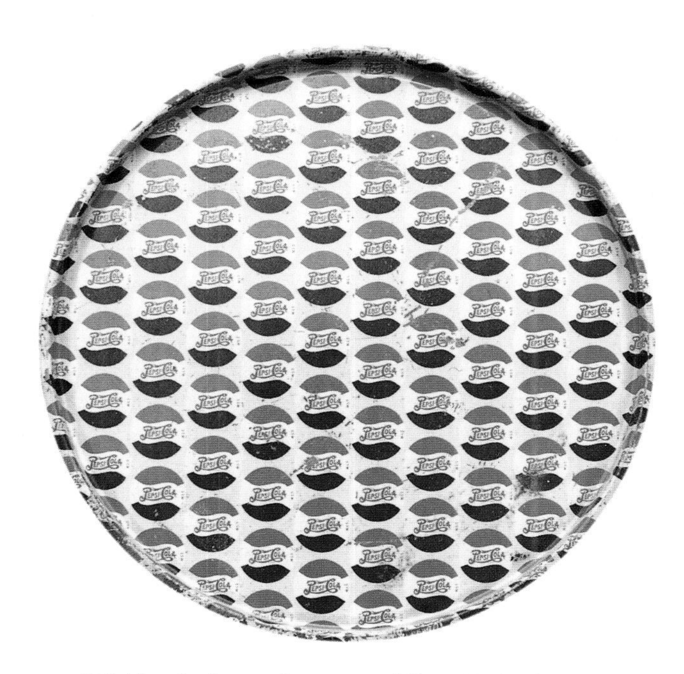

668 Metal, sheet of crowns, 15", c.1940s, $100.00

670 Metal, 12", c.1960s, $75.00

671 Plastic, Tiffany style, 13", c.1970s, $30.00

672 Plastic, Slice, 13", c.1980s, $20.00

Umbrellas

673 Nylon, c.1970s, $25.00

676 Cloth, c.1980s, $80.00

674 Nylon, c.1980s, $25.00

675 Nylon, c.1980s, $10.00

677 Vinyl, c.1990s, $80.00

Vending Machines

679 Electric, water cooled, c.1960s, $250.00

678 Electric, bottle type, Jacobs 56, c.1950s, $4,500.00

681 Electric, air cooled, c.1960s, $200.00

680 Electric, can type, c.1960s, $300.00

Watches

682 Pepsi Light, lemon second hand, c.1970s, $40.00

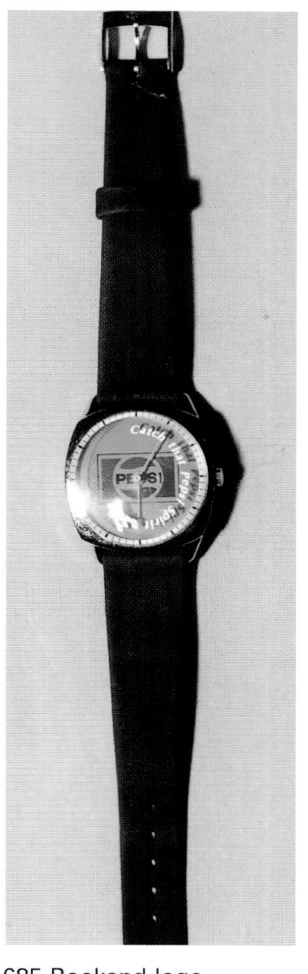

685 Bookend logo, bi-plane second hand, c.1980s, $40.00

686 Teem, c.1960s, $50.00

684 Bookend logo, bottle second hand, c.1980s, $40.00

683 Pocket watch with case, c.1980s, $55.00

687 Quartz, in plastic cans, c.1980s, $25.00 each

689 Truck second hand, c.1980s, $40.00

688 Pepsi 400, 1992, $125.00

690 Set of five with display box, 1996, $125.00

MORE SCHIFFER TITLES

www.schifferbooks.com

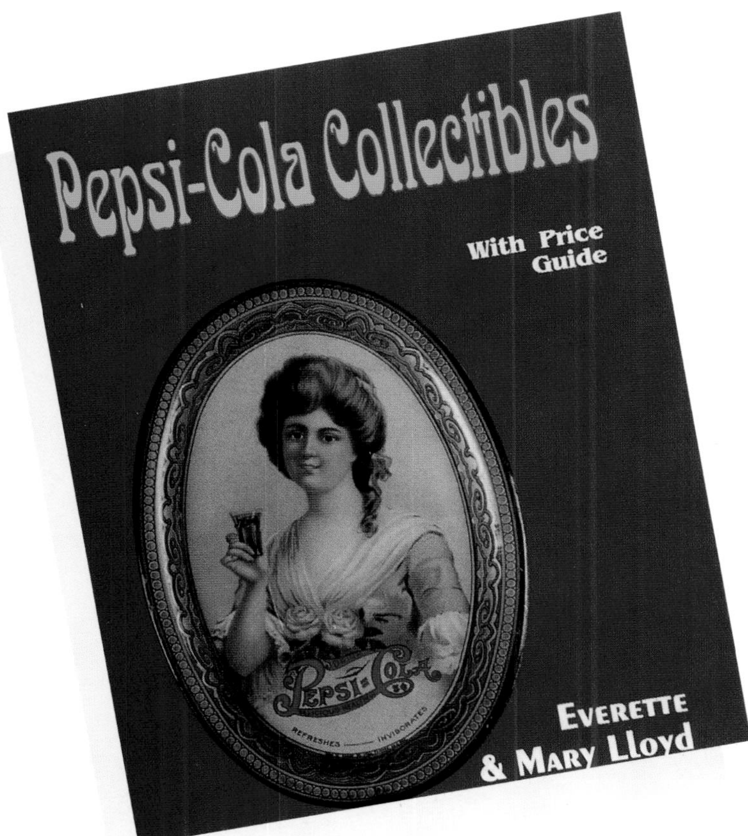

Pepsi-Cola Collectibles. Everette and MaryLloyd. In the late 1890s, Caleb Davis Bradham, a North Carolina drugstore owner and pharmacist, mixed a concoction at the soda fountain for his customers that was a hit. They called it "Brad's Drink"; Bradham named the mixture Pepsi Cola. Pepsi became a national success and engendered an array of commercial, advertising, and promotional objects that are much sought after by collectors today. Many of those objects are included in this book, presented with beautiful, full-color photographs. The items come from one of the largest collections of Pepsi memorabilia in the nation, that of Everette Lloyd.
Size: 8 1/2" x 11" 400 color photos 160 pp.
Price Guide
ISBN: 0-88740-533-9 soft cover $29.95

Just for Openers: A Price Guide to Beer, Soda & Other Openers. Donald A. Bull & John R. Stanley. Advertising bottle openers and corkscrews made exclusively for brewing, malt, and soda companies in the U. S. are studied and shown. Coca-Cola, Pepsi-Cola, Dr. Pepper, and Nehi soft drinks are among the many beverages advertised on bottle openers. Grouped by category and alphabetically, over 850 remarkable advertising openers appear with descriptions, history and current values. An index and list of American patents make this a valuable reference that may "open" a fascinating new hobby.
Size: 8 1/2" x 11" 382 color photos 160 pp.
Price Guide Index
ISBN: 0-7643-0846-7 soft cover $29.95

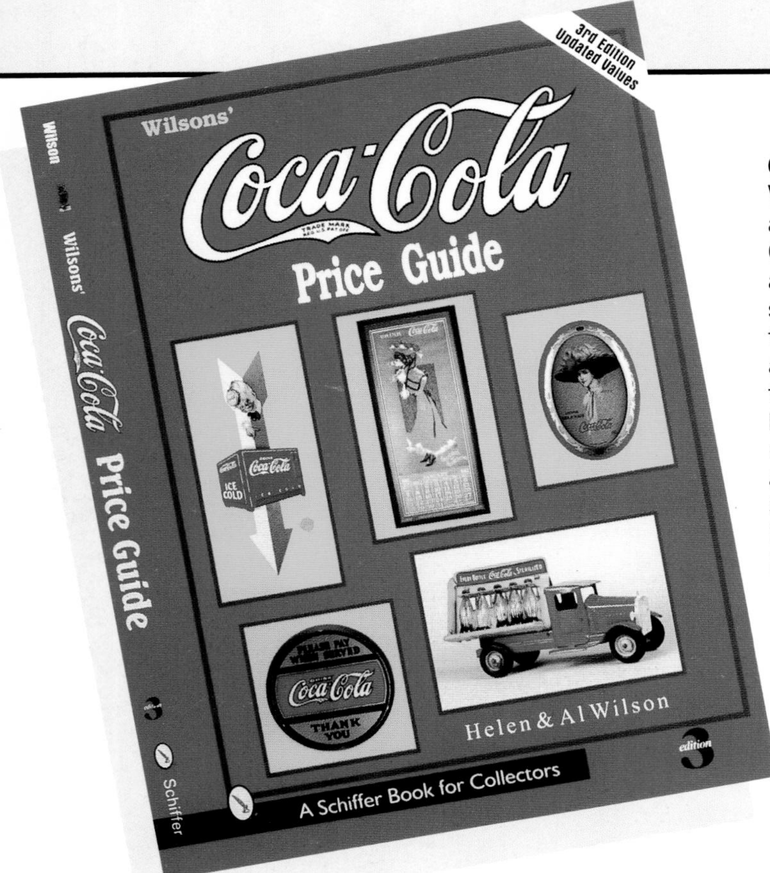

Coca-Cola Price Guide. *Revised 3rd Edition*. Al and Helen Wilson. Now in soft cover, this is a must-have reference for anyone interested in the wondrously broad range of Coca-Cola collectibles! This book is filled cover-to-cover with approximately 2,000 bright, color pictures of the most sought-after Coca-Cola products ever made, from advertisements, trays, and bottles to haberdashery, jewelry, and amazing one-of-a-kind novelties. From its earliest offerings to a modern-day selection, Coca-Cola is well represented by this definitive new text, which includes many of the rarest pieces from within the Coca-Cola Company's own archives. A useful and gloriously beautiful book!
Size: 9" x 12" 2000 photos 256 pp.
Price Guide
ISBN: 0-7643-0983-8 soft cover $29.95

Coca-Cola Trays. *Revised and Expanded 2nd Edition.* William McClintock. Ever since 1897, when the first "Delicious and Refreshing" glasses of Coca-Cola were poured, Coca-Cola trays have been made as magnificent pieces of advertising history, showing the evolution of American popular culture. From ribbons-and-lace girls of the late Victorian era through Roaring Twenties flappers, World War II war brides, and the working women of today, Coca-Cola has called upon images of glamour girls and girls-next-door to sell Coke. Warm family scenes, baseball, and children at play are also favorite tray illustrations. A price guide is included.Through the 250 trays shown in color, you will come to understand the charm and appeal of these trays.
Size: 6" x 9" 250 color photos 144 pp.
Revised Price Guide
ISBN: 0-7643-0984-6 soft cover $12.95